Steel Fire Designer's Handbook

with Worked Examples

Nominal Fire Resistance according to EUROCODES

Primoz Kvaternik, M.Eng.

GRSOFT Eurocode Designer's Handbooks

GrSoft

Published by Primoz Kvaternik, GRSOFT Structural Engineering
http://www.grsoft.eu

First published 2014

ISBN: 978-961-281-433-5

You can contact us for consultancy at:
http://www.grsoft.eu/   engs@grsoft.eu

Blog on Fire Analysis
http://www.fire-analysis-design.com/

Other Amazon Titles
Engineering Calculations using Microsoft Excel

Udemy Courses
Fire Analysis of Steel Structures according to Eurocodes
Engineering Calculations using Microsoft Excel

Contents

PREFACE

Fire resistance of a building structure means the ability of the structure exposed to fire maintaining its strength for the appropriate time. To recognize the precise value of the duration in which the construction retains its capacity, is crucial for fire-resistant design. Fire analysis calculation takes into account the actual technical state of construction material exposed to thermal impact or fire load and provide you with the critical temperature, e.g. the temperature to which the structure maintains its capacity or advice you of the optimal fire protection of the structural elements.

Why must we learn Fire Analysis?

Because we do not want that we have fire protection, which is oversized and as well uneconomical. On the other hand we do not want that the construction is in fire risk and does not allow a safe evacuation, as well as fire extinguishing and rescue.

In this book I will present a thorough overview of structural fire analysis according to nominal temperature-time curves according to Eurocodes.

We will learn the following topics:

- What is thermal action?
- How to calculate temperature of protected and unprotected steel sections
- How to create load combinations for the fire limit states
- How to calculate fire resistance of structural members
- How to calculate critical temperature
- Fire protection systems for steel structures
- Worked example of real structure

All chapters will also provide you with worked examples which will cover theoretical part presented. At the end you will be presented a calculation of critical temperature and required fire protection for typical structural elements of real structure.

Primoz Kvaternik, M.Eng.

ABOUT THE AUTHOR

Primoz Kvaternik, M.Eng.

Primoz is a director of **GRSOFT Structural Engineering Company** in Slovenia / EU. As a Structural Engineer with more than 15 years' experience in structural design on numerous architectural-engineering building projects both large and small, he has been working on project documentation for all types of building structures, reinforcement plans for RC structures, fabrication plans of steel structures, fire resistance calculation of building structures including consultation on the most appropriate fire protection. Based on years of experience and participation in post-earthquake reconstruction, he specialize in Earthquake Reconstruction and rehabilitation of buildings as well.

He has university degree in structural engineering and all possible licenses for the design and revision of project documentation for building structures at **Slovenian Chamber of Engineers**.

He is currently teaching how to use Eurocodes as one of the most advanced building design codes on a world scale. He is teaching also how to use technology for solving daily engineering calculations, including software development as a support to structural engineering.

1

IMPACT OF FIRE ON STRUCTURE

In this Chapter

- Thermal action
- Heat transfer mechanisms
- Section factor
- Temperature of unprotected steel sections
- Temperature of protected steel sections

Thermal Action

What is thermal action and how we consider thermal action on the structure?

According to Eurocodes thermal action is defined as temperature-time curve, which represents time evolution of a gas temperature surrounding the structure. On the basis of this temperature we can calculate the heat flux which is transmitted from surrounding to the structure.

We are talking about nominal temperature-time curve as it does not represent temperature of real fire, but instead measured temperature of previous fires and is as such standardized.

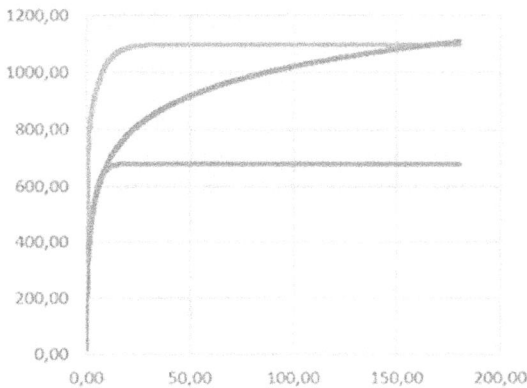

Fig. 1.1 Nominal temperature-time curves

Eurocode 1 (EN 1991-2-2) introduces three different nominal temperature-time curves.

- Standard or ISO curve
- External curve
- Hydrocarbon curve

The most commonly used is so called standard or ISO curve. Then we have external curve, which is exactly the same as standard curve till something more than 600 degrees Celsius, then remains constant. At last we have hydrocarbon curve.

Each of this curves is used for special types of fires. Let's look these curves in detail.

Standard or ISO curve

It is most commonly used in standard fire test for evaluation of structural elements or separating walls. It is used when fire is not localized, but instead fully developed in fire compartment. Here we can see the equation for this curve, where the only parameter is time, which means the time of fire exposure and should be entered in minutes.

$$T = 20 + 345 log_{10}(8t + 1) ... \ (1.1)$$

The curve means the temperature of gasses surrounding the structural element or separating wall in degrees Celsius.

External curve

It is used for external walls exposed to fire or for outer surface of separating walls of fire compartment. Here we can see slightly different equation for this curve, where again the only parameter is time, which means the time of fire exposure and should be entered in minutes. The curve means the temperature of gasses surrounding the external or separating walls.

$$T = 20 + 660(1 - 0.687e^{-0.32t} - 0.313e^{-3.8t}) ... \ (1.2)$$

Hydrocarbon curve

It is used for full combustion of hydrocarbons, for instance gasoline, oil or other fuels. We can see the equation for this curve below, where is again the only parameter time. The curve means the temperature of gasses surrounding the hydrocarbons in fire compartment.

$$T = 20 + 1080(1 - 0.325e^{-0.167t} - 0.675e^{-2.5t}) ... \ (1.3)$$

Example 1.1

Let's see the practical example where we will calculate the temperature of all three curves after 30 minutes of fire exposure.

What will be the temperature of all three nominal curves after 30 minutes?

Standard curve

$$T = 20 + 345 log_{10}(8 * 30 + 1) = 841.8°C$$

External curve

$$T = 20 + 660(1 - 0.687e^{-0.32*30} - 0.313e^{-3.8*30}) = 679.97°C$$

Hydrocarbon curve

$$T = 20 + 1080(1 - 0.325e^{-0.167*30} - 0.675e^{-2.5*30}) = 1097.66°C$$

Heat Transfer Mechanisms

Here we will learn how the heat from surrounding of the steel element is transferred to the element.

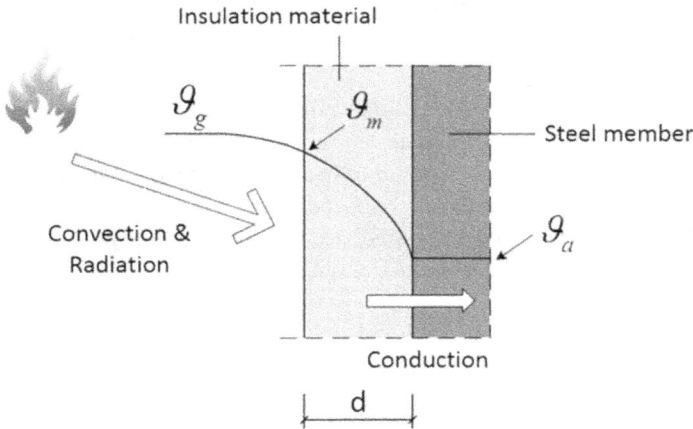

Fig. 1.2 Heat transfer mechanisms

Increased temperature of the steel section is dependent on the area of the steel exposed to fire, from the amount of fire protection and on the temperature of the fire compartment. The heat of the gasses from the fire compartment is transferred to the surface of the insulation or steel element by convection and radiation. Through the steel element is heat transferred by conduction

Section Factor

Let's learn first what is section factor?

Section factor is a rate of heating of cross section and is defined as the area of the steel section exposed to fire divided by the cross section of the steel element.

We need to know here one important fact that the cross sections with higher section factors heat faster than the ones with a lower one.

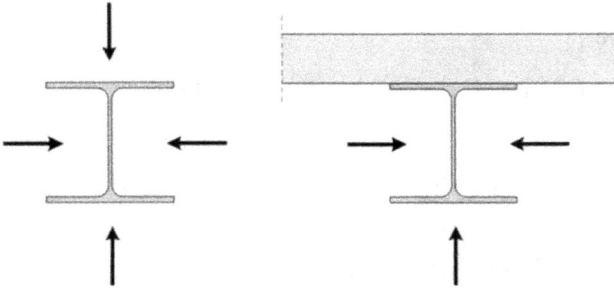

Fig. 1.3 Section factor for unprotected steel cross sections

$$\frac{A_m}{V} = \frac{Steel\ perimeter}{Cross\ section\ area} \ ...\ (1.4)$$

In the picture above we can see calculation of the section factor of the unprotected steel sections. Steel perimeter means the area of the steel section exposed to fire. In the left side of the picture cross section is exposed to fire on all four sides, when on the right side of the picture it is exposed to fire just on three sides as it supports concrete slab. It is obvious that the section factor for the steel section supporting slab is lower as it heats slower as it is the case for the steel section alone.

Let's see the calculation of section factor for protected steel sections now.

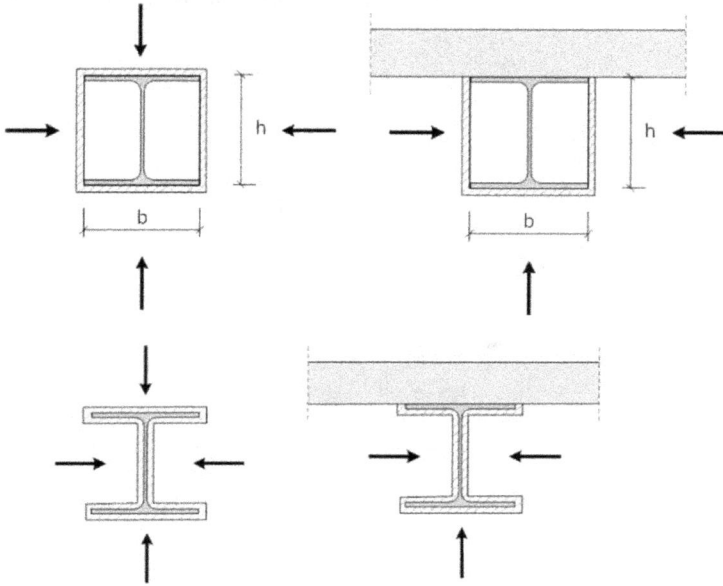

Fig. 1.4 Section factor for protected steel cross sections

$$\frac{A_p}{V} = \frac{Encasement\ perimeter}{Cross\ section\ area}\ ...(1.5)$$

Encasement perimeter is different if we have contour or hollow encasement. Again section factor is lower if the steel beam is supporting concrete slab as cross section is exposed to fire just from three sides.

Example 1.2

Let's calculate a section factor for unprotected steel section HEA-240 exposed to fire from all 4 sides.

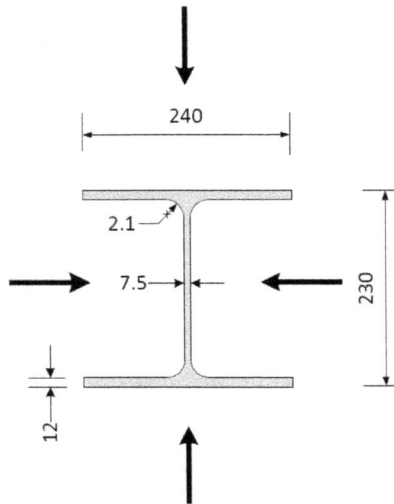

Fig. 1.5 Cross section dimensions for unprotected HEA-240 profile

First we need to calculate a cross section area of steel section, then we will calculate the perimeter of steel section exposed to fire and finally the section factor as the ratio of these two values.

$$V = 76{,}8 \ cm^2 \ ... \ \text{HEA-240}$$

$$A_m = 24 * 2 + 1.2 * 4 + 4 * \frac{(24 - 0{,}75 - 2.1 * 2)}{2} + 2 * \pi * 2.1 + 2$$
$$* (23 - 2 * 1.2 - 2 * 2.1) = 136.89 \ cm$$

$$\frac{A_m}{V} = \frac{136.89 \ cm}{76.8 \ cm^2} = 178.24 \ m^{-1}$$

Example 1.3

Let's calculate a section factor for protected steel section HEA-240, supporting a concrete slab, exposed to fire from three sides.

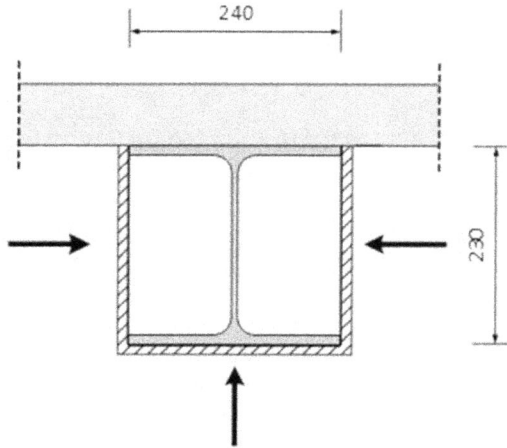

Fig. 1.6 Cross section dimensions for protected HEA-240 profile

Again we need to calculate a cross section area of steel section, then we will calculate the encasement perimeter of the insulation of steel section exposed to fire and finally the section factor as the ratio of these two values.

$$V = 76{,}8 \; cm^2 \; ... \; \text{HEA-240}$$

$$A_p = 2 * 23 + 24 = 70 \; cm$$

$$\frac{A_p}{V} = \frac{70 \; cm}{76.8 \; cm^2} = 91.1 \; m^{-1}$$

Temperature of Unprotected Steel Sections

Eurocode 3, second part (EN 1993-1-2) introduces the following formula for the calculation of the increase in temperature of the steel element exposed to fire in time interval.

$$\Delta\vartheta_{a,t} = k_{sh}\frac{\frac{A_m}{V}}{c_a\rho_a}h_{net,d}\,\Delta t\ ...\,(1.6)$$

A_m/V	section factor for unprotected steel section [m^{-1}]
k_{sh}	correction factor for the shadow effect
ρ_a	unit mass of steel [kg/m^3]
c_a	specific heat of steel [J/kgK]
$h_{net,d}$	net heat flux [W/m^2]
Δt	time interval $\leq$ 5s [s]

Net heat flux is defined in Eurocode 1, second part (EN 1991-1-2) where the following formula is introduced. We can calculate the convective and radiative part of it at current temperature of iteration.

$$h_{net,d} = \alpha_c\left(\vartheta_g - \vartheta_m\right) + \Phi\varepsilon_m\varepsilon_f\sigma\left[(\vartheta_r + 273)^4 - (\vartheta_m + 273)^4\right]\ ...\,(1.7)$$

α_c	convection heat transfer coefficient [W/m^2 K]
ϑ_g	gas temperature of the fire compartment [˚C]
ϑ_m	temperature of the surface of the steel element [˚C]
ϑ_r	radiation temperature of the fire [˚C]
Φ	view factor
ε_m	emisivity of the steel
ε_f	emisivity of the flame
σ	Stephan Boltzman constant = 5.67x10^{-8} [W/m^2 K^4]

View factor Φ is normally taken as 1.0, lower value can be used when the shadow effect is active or when the fire is localized (EN 1992-1-2: Annex G)

Shadow effect is active at the concave shapes of the profile as seen bellow.

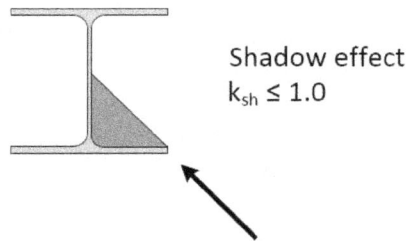

Shadow effect
$k_{sh} \leq 1.0$

Fig. 1.7 Shadow effect for concave shapes of cross sections

For I sections under nominal fire we can calculate the correction coefficient for shadow effect using the formula bellow.

$$k_{sh} = 0.9 \frac{\left[\frac{A_m}{V}\right]_b}{\left[\frac{A_m}{V}\right]} \dots (1.8)$$

Where section factor with the mark b means box value of the section factor. In all other cases, we can calculate the correction coefficient for shadow effect using the formula bellow.

$$k_{sh} = \frac{\left[\frac{A_m}{V}\right]_b}{\left[\frac{A_m}{V}\right]} \dots (1.9)$$

Shadow effect is not active at the convex shapes of the profile as seen bellow.

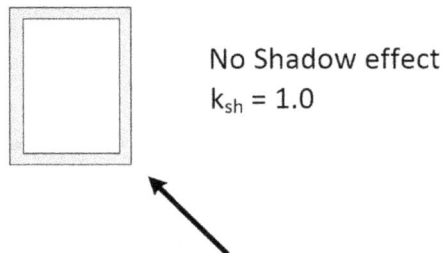

No Shadow effect
$k_{sh} = 1.0$

Fig. 1.8 Shadow effect for convex shapes of cross sections

Convection heat transfer coefficient α_c is defined in Eurocode 1, second part (EN 1991-1-2 (3.2.1)) and is dependent on which nominal temperature-time curve we use.

Radiation temperature can be taken as the gas temperature of the fire compartment for fully flame engulfed structural element.

Emissivity of the flame and emissivity of the steel are defined in Eurocode 1, second part (EN 1991-1-2) and means the ability to emit heat by radiation.

To calculate the heat transfer by using the equation presented above we must use the iterative procedure, where in each iteration we need to calculate temperature dependent parameters as specific heat of steel, net heat flux and others.

Example 1.4

Let's show the application of the formulas presented and calculate a temperature of unprotected steel section HEA-240 after 30 minutes of standard fire exposure on all four sides.

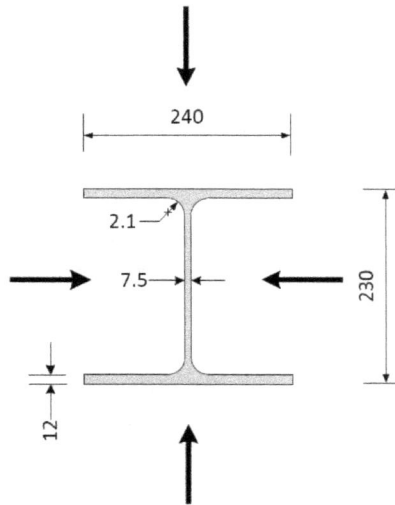

Fig. 1.9 Cross section dimensions for unprotected HEA-240 profile

At the starting moment t=0s, the steel temperature is equal to the room temperature 20°C and we have

$$\vartheta_m = 20°C, t = 0s$$

The section factor, box value of the section factor and the correction factor for the shadow effect are calculated as follows

$$\frac{A_m}{V} = 178.24 \ m^{-1} \quad \left[\frac{A_m}{V}\right]_b = \frac{94}{76.8} * 100 = 122.4 \ m^{-1}$$

$$k_{sh} = 0.9 * \frac{122.4}{178.24} = 0.62$$

First time interval t_1 can be taken as follows

$$\Delta t = 5s \longrightarrow t_1 = 0 + \Delta t = 5s$$

Gas temperature of the fire compartment is calculated using the standard temperature-time curve

$$\vartheta_g = 20 + 345 \log_{10}\left(\frac{8t_1}{60} + 1\right) = 96.54°C$$

Specific heat of steel is temperature dependent and is calculated to EN 1993-1-2(3.4.1.2) as follows

$$c_a = 425 + 0.773\vartheta_m - 0.00169\vartheta_m^2 + 0.00000\vartheta_m^3 = 439.8 \, J/kgK$$

Net heat flux can be calculated to EN 1991-1-2(3.1)

$$h_{net,d} = \alpha_c(\vartheta_g - \vartheta_m) + \Phi\varepsilon_m\varepsilon_f\sigma[(\vartheta_r + 273)^4 - (\vartheta_m + 273)^4]$$
$$= 2361.07 \, W/m^2$$

Using EN 1991-1-2(3.2.1, 3.1(6, 7)) we have the following parameters used in the formula above

$$\alpha_c = 25 \, W/m^2 \, K \quad \varepsilon_m = 0.7 \quad \varepsilon_f = 1.0 \quad \Phi = 1.0$$

$$\sigma = 5.67 * 10^{-8} \, W/m^2 \, K^4$$

For fully fire engulfed steel members we can take

$$\vartheta_{r=}\vartheta_g$$

The increase in steel temperature of the member in the time interval can be calculated as follows

$$\Delta\vartheta_{a,t} = k_{sh}\frac{\frac{A_m}{V}}{c_a\rho_a}h_{net,d}\,\Delta t = 0.62 * \frac{178.24}{439.8 * 7850} * 2361.1 * 5.0$$
$$= 0.378°C$$

Temperature in the first interval can be calculated as follows

$$\vartheta_m = \vartheta_m + \Delta\vartheta_{a,t} = 20 + 0.378 = 20.378°C$$

The next interval can be defined as

$$t_2 = t_1 + \Delta t = 5 + 5 = 10s$$

Now we can calculate the increase in temperature of the steel member in the next interval as follows

$$\vartheta_g = 146.95°C \quad c_a = 440.07 \, J/kgK \quad h_{net,d} = 4104.83W/m^2$$

$$\Delta\vartheta_{a,t} = 0.655°C \quad \vartheta_m = 20.378 + 0.655 = 21.03°C$$

The next interval can be defined as

$$t_i = t_{i-1} + \Delta t = 10 + 5 = 15s$$

The other iterations of calculation are presented in the following table:

t [min]	5	10	15	20	25	30
$\vartheta_g[°C]$	576.40	678.40	738.56	781.35	814.60	841.80
$\vartheta_a[°C]$	193.30	418.85	588.92	692.68	735.90	782.26

Temperature of the steel section after 30 minutes of standard fire exposure is 782.26 °C.

Temperature of protected steel sections

Eurocode 3, second part (EN 1993-1-2) proposes for steel sections with applied thermal insulation and uniform temperature distribution throughout the cross section the following formula for the calculation of the increase in temperature of the steel element exposed to fire in time interval.

$$\Delta\vartheta_{a,t} = \frac{\lambda_p A_p/V (\vartheta_{g,t} - \vartheta_{a,t})}{d_p c_a \rho_a (1 + \Phi/3)} \Delta t - \left(e^{\Phi/10} - 1\right)\Delta\vartheta_{g,t} \ldots (1.10)$$

Where the following parameter means the amount of heat stored in insulation

$$\Phi = \frac{c_p d_p \rho_p}{c_a \rho_a} \frac{A_p}{V}$$

A_p/V	section factor for protected steel section [m^{-1}]
λ_p	thermal conductivity of fire protection [W/mK],
$\vartheta_{g,t}$	gas temperature of the fire compartment [°C]
$\vartheta_{a,t}$	temperature of the surface of the steel element [°C]
Δt	time interval $\leq$ 30s [s]
ρ_a	unit mass of steel [kg/m^3]
ρ_p	unit mass of insulation [kg/m^3]
c_a	specific heat of steel [J/kgK]
c_p	specific heat of insulation material [J/kgK]
d_p	depth of fire insulation material [m]

Limitations of the upper equation

- According to ECCS Eurocode Design Manual, equation is only valid for small values of the factor $\Phi \leq 1.5$ [Ref 10]
- Thermal resistance against heat conduction is achieved only by fire protection and not from the steel element
- According to Eurocodes fire protection material properties should be tested experimentally
- Thermal resistance between gasses and the outer surface of the fire protection is neglected
- Heat transfer by convection and radiation is not covered at all
- Thermal conductivity of insulation at elevated temperature should not be used the same as for room temperature

The Bottom Line of the facts presented above is that heat transfer of fire protected steel elements is a complex analysis and that the use of simple formula defined in EN 1993-1-2 should be taken very carefully. More complete information can be found in ECCS Eurocode Design Manual [Ref 10]

Example 1.5

Let's show the application of the formulas presented and calculate a temperature of protected steel section HEA-240 supporting a concrete slab after 30 minutes of standard fire exposure.

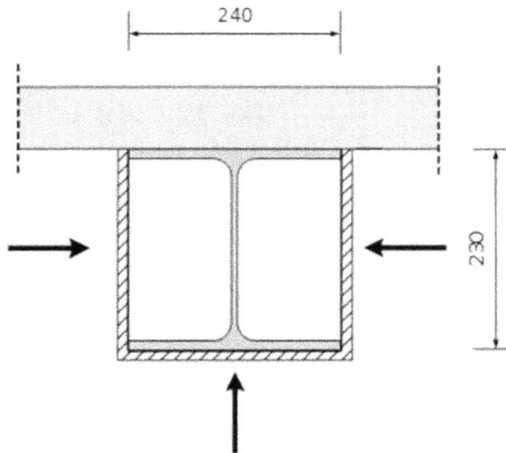

Fig. 1.10 Cross section dimensions for protected HEA-240 profile

Steel member is protected with fibre-silicate boards with the following parameters

$$\rho_p = 600\,\frac{kg}{m^3} \quad c_p = 1200\,\frac{J}{kgK} \quad \lambda_p = \frac{0.15W}{mK} \quad d_p = 0.02m$$

The section factor and the unit mass of steel are calculated as follows

$$\frac{A_p}{V} = \frac{70\,cm}{76.8\,cm^2} = 91.1\,m^{-1} \quad \rho_a = 7850\,\frac{kg}{m^3}$$

First time interval t_1 can be taken as follows

$$\Delta t = 30s \longrightarrow t_1 = 0 + \Delta t = 30s$$

Gas temperature of the fire compartment is calculated using the standard temperature-time curve

$$\vartheta_{g,t} = 20 + 345 \log_{10}\left(\frac{8t_1}{60} + 1\right) = 261.14°C$$

The increase in gas temperature of the fire compartment is calculated as

$$\Delta\vartheta_{g,t} = 261.14 - 20 = 241.14°C$$

Specific heat of steel is temperature dependent and is calculated to EN 1993-1-2(3.4.1.2) as follows

$$\vartheta_{a,t} = 20°C \quad \rightarrow c_a = 425 + 0.773\vartheta_{a,t} - 0.00169\vartheta_{a,t}^2 + 0.000000\vartheta_{a,t}^3$$
$$= 439.8\,J/kgK$$

The amount of heat stored in the insulation is calculated as

$$\Phi = \frac{c_p d_p \rho_p}{c_a \rho_a}\frac{A_p}{V} = \frac{1200 * 600 * 0.02}{439.8 * 7850} * 91.1 = 0.38$$

The increase in steel temperature of the member in the time interval can be calculated as follows

$$\Delta\vartheta_{a,t} = \frac{\lambda_p A_p/V(\vartheta_{g,t} - \vartheta_{a,t})}{d_p c_a \rho_a(1 + \Phi/3)}\Delta t - \left(e^{\frac{\Phi}{10}} - 1\right)\Delta\vartheta_{g,t}$$

$$= \frac{0.15 * 91.1 * 241.14 * 30}{0.02 * 439.8 * 7850 * \left(1 + \dfrac{0.38}{3}\right)} - \left(e^{\frac{0.38}{10}} - 1\right)$$

$$* 241.14 = -8.06°C$$

As negative increments of steel temperature are not allowed, we have

$$\Delta\vartheta_{g,t} = 241.14°C \quad \rightarrow \Delta\vartheta_{a,t} = 0°C$$

Second time interval t_2 can be taken as follows

$$t_2 = t_1 + \Delta t = 30 + 30 = 60s$$

Gas temperature of the fire compartment is calculated using the standard temperature-time curve

$$\vartheta_{g,t} = 20 + 345 \, log_{10}\left(\frac{8t_2}{60} + 1\right) = 349.21°C$$

The increase in gas temperature of the fire compartment is calculated as

$$\Delta\vartheta_{g,t} = 349.21 - 261.14 = 88.07°C$$

Specific heat of steel and the amount of heat stored in the insulation are the same as at the previous iteration

$$\vartheta_{a,t} = 20°C \quad \rightarrow c_a = 425 + 0.773\vartheta_{a,t} - 0.00169\vartheta_{a,t}^2 + 0.00000\vartheta_{a,t}^3$$

$$= 439.8 \frac{J}{kgK} \quad \Phi = 0.38$$

The increase in steel temperature of the member in the time interval can be calculated as follows

$$\Delta\vartheta_{a,t} = \frac{\lambda_p A_p/V\left(\vartheta_{g,t} - \vartheta_{a,t}\right)}{d_p c_a \rho_a(1 + \Phi/3)}\Delta t - \left(e^{\frac{\Phi}{10}} - 1\right)\Delta\vartheta_{g,t}$$

$$= \frac{0.15 * 91.1 * 329.21 * 30}{0.02 * 439.8 * 7850 * \left(1 + \frac{0.38}{3}\right)} - \left(e^{\frac{0.38}{10}} - 1\right)$$

$$* 88.07 = -1.67°C$$

The other iterations of calculation are presented in the following table:

t [min]	5	10	15	20	25	30
$\vartheta_g[°C]$	576.40	678.40	738.56	781.35	814.60	841.80
$\vartheta_a[°C]$	31.50	58.10	87.70	118.00	148.17	177.86

Temperature of the steel section after 30 minutes of standard fire exposure is 177.86°C.

2

FIRE DESIGN FUNDAMENTALS

In this Chapter

- Load combinations
- Introduction to mechanical analysis
- Classification of cross sections

Load Combinations

According to Eurocodes the fire load combinations are defined in EN 1990 and are different as the ones for the room temperature. In general, load factors are lower as for room temperature, because the fire is more rare situation event that in normal conditions. According to EN 1990 fire load combination is accidental situation.

Design action effects for the fire load situation can be according to EN 1990, calculated using the following expression

$$E_{fi,d,t} = \sum_{j \geq 1} G_{k,j} + P + A_d + \psi_{1,1} Q_{k,1} + \sum_{i \geq 2} \psi_{2,i} Q_{k,i} \ldots (2.1)$$

$G_{k,j}$ permanent actions

P prestressing actions

A_d indirect fire actions

$Q_{k,1}$ leading variable action

$Q_{k,i}$ accompanying variable action

As we can see here, permanent actions are used with characteristic values without load factors. This is also the case for pre-stressing and indirect fire actions. When variable actions have load factors which are much lower than the ones for the room temperature.

The formula presented above and load factors can be slightly different at individual EU member states. Check for national annexes.

Action	ψ_1	ψ_2
Imposed loads in buildings		
Category A: domestic, residental areas	0,5	0,3
Category B: office areas	0,5	0,3
Category C: congregation areas	0,7	0,6
Category D: shopping areas	0,7	0,6
Category E: storage areas	0,9	0,8
Category F: traffic area, vehicle weight ≤ 30kN	0,7	0,6
Category G: traffic area, 30kN≤vehicle weight ≤ 160 kN	0,5	0,3
Snow loads on buildings		
Finland, Iceland, Norway, Sweden	0,5	0,2
Reminder of countries, H > 1000m	0,5	0,2
Reminder of countries, H ≤ 1000m	0,2	0,0
Wind loads on buildings	0,2	0,0

Fire resistance verification

If the effects of actions can be done normally, this is usually the case when we create separate structural model for fire analysis, we can use general rule

$$E_{fi,d,t} \leq R_{fi,d,t} \dots (2.2)$$

$E_{fi,d,t}$ effects of actions at fire load combination

$R_{fi,d,t}$ fire resistance of structural members at elevated temperature

If the effects of actions are already calculated in the structural model for the room temperature, it is easier to use simplified rules, where the effects of actions in fire load combination are calculated from the effects of actions at room temperature

$$E_{fi,d,t} = \eta_{fi}E_d \quad \eta_{fi} = \frac{G_k + \psi_{fi}Q_{k,1}}{\gamma_G G_k + \gamma_{Q,1}Q_{k,1}} \dots (2.3)$$

ψ_{fi} combination factor for fire load combination

γ_G load factor for permanent actions at room temperature

$\gamma_{Q,1}$ load factor for variable actions at room temperature

G_k permanent action

$Q_{k,1}$ leading variable action

Example 2.1

Let's show the application of these rules in the following example. We have simply supported beam which is located at congregation area of type C2. Beam has a span of 8m and is subjected to permanent load of 15kN/m and live load of 20kN/m. We need to calculate the design bending moment in fire situation!

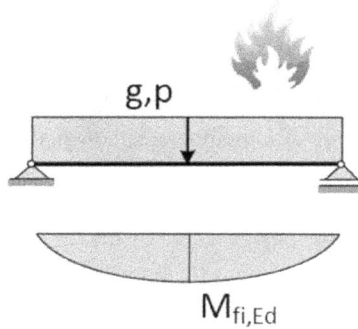

Fig. 2.1 Simply supported beam subjected to uniform load

Using general rule for fire resistance we have

$$q_{fi,Ed} = 15 + 0.7 * 20 = 29kN/m$$

$$M_{fi,Ed} = \frac{29 * 8^2}{8} = 232.0kNm$$

Using simplified rule for fire resistance we have

$$q_{fi,Ed} = 1.35 * 15 + 1.5 * 20 = 50.25kN/m$$

$$M_{Ed} = \frac{50.25 * 8^2}{8} = 402.0kNm$$

$$\eta_{fi} = \frac{15 + 0.7 * 20}{1.35 * 15 + 1.5 * 20} = 0.577$$

$$M_{fi,Ed} = 0.577 * 402 = 231.95kNm$$

We need to be aware that simplified rules always give us an approximate values which are not always so close as in this example. If it is possible it is advisable to use general rule.

Introduction to Mechanical Analysis

Verification of the fire resistance should be made according to EN 1993-1-2 by using the same time interval as used in temperature analysis. The following two domains are suggested.

Strength domain

For the required duration of time the effects of actions in fire situation should be less than or equal to the fire resistance of structural members at elevated temperature

$$t_{fi,req} \rightarrow E_{fi,d,t} \leq R_{fi,d,t} \dots (2.4)$$

$E_{fi,d,t}$ effects of actions at fire load combination

$R_{fi,d,t}$ fire resistance of structural members at elevated temperature

Temperature domain

For required duration of time the design value of the temperature of structural member in fire compartment should be less than or equal to design value of member critical temperature

$$t_{fi,req} \rightarrow \vartheta_{d,t} \leq \vartheta_{d,cr} \dots (2.5)$$

$\vartheta_{d,t}$ design value of member temperature

$\vartheta_{d,cr}$ design value of member critical temperature

For nominal fire resistance member analysis should be applied on isolated members extracted from the whole structure. The effects of actions calculated at room temperature can be used also at elevated temperature.

Reduction factors at elevated temperatures

In the table and chart below we can see reduction factors for yield strength and Young's modulus at elevated temperatures.

$\vartheta_a\,[°C]$	$k_{y,\vartheta}$	$k_{E,\vartheta}$
20	1.000	1.000
100	1.000	1.000
200	1.000	0.900
300	1.000	0.800
400	1.000	0.700
500	0.780	0.600
600	0.470	0.310
700	0.230	0.130
800	0.110	0.090
900	0.060	0.067
1000	0.040	0.045
1100	0.020	0.022
1200	0.000	0.000

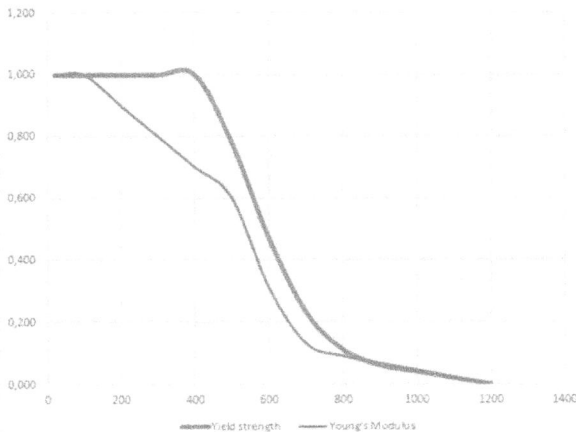

Fig. 2.2 Yield strength and elastic modulus at elevated temperatures

As we can see, the yield strength has at 600 degrees celzius just 47% of it's value at room temperature and elastic modul only 31% of it's value at room temperature.

Classification of Cross Sections

To prevent local buckling of individual parts of steel cross sections, loded in compression and bending, the ratio length to the width of plate elements should be limited. That way it is prevented that colapse occurs before section reaches its yield strength.

EN 1993-1-1 defines four classes of cross sections:

- **Class 1** cross sections can form a plastic hinge with the rotation capacity required without reducing the cross section resistance.

- **Class 2** cross sections can develop their plastic moment resistance, but have limited rotation capacity because of local buckling effects.

- **Class 3** cross sections are those where calculated stresses in the extreme compression fibre can reach its yield strength, but local buckling is liable to prevent development of the plastic moment resistance of the cross section.

- **Class 4** cross sections are those where local buckling will occur before reaching the yield strength in one or more parts of the cross section.

We will not go here in details of cross section classification, as it is part of designing of steel structures at room temperature and is defined in EC 3 first part (EN 1993-1-1).

For the purpose of Nominal Fire Analysis, cross section classification can be done on the same way as for normal temperature, but with a reduced value of the parameter of section classification as follows

$$\varepsilon = 0.85 \sqrt{\frac{235}{f_y}} \dots (2.6)$$

f_y yield strength of steel

Example 2.2

Calculate the cross section class for HEA-240 in steel grade S235 exposed to fire. Consider simply supported beam with pure bending!

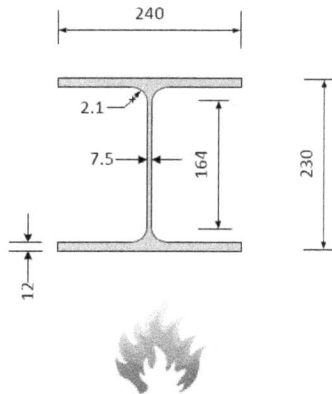

Fig. 2.3 Cross section dimensions for unprotected HEA-240 profile

Reduced value of the parameter of section classification for steel grade S235

$$\varepsilon = 0.85 \sqrt{\frac{235}{f_y}} = 0.85 \sqrt{\frac{235}{235}} = 0.85$$

The class of the web in bending

$$\frac{c}{t_w} = \frac{16.4}{0.75} = 21.87 < 72\varepsilon = 61.2 \rightarrow Class\ 1$$

The class of the flange in compression

$$\frac{c}{t_f} = \frac{9.52}{1.20} = 7.94 < 10\varepsilon = 8.5 \rightarrow Class\ 2$$

The cross section is Class 2!

Example 2.3

Calculate the cross section class for HEA-240 in steel grade S235 exposed to fire. Consider simply supported column with pure compression!

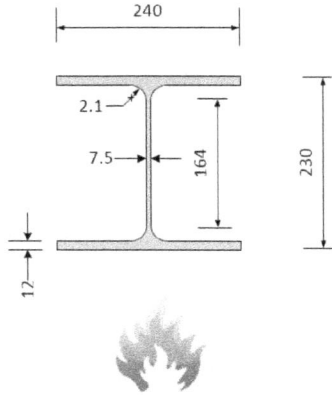

Fig. 2.4 Cross section dimensions for unprotected HEA-240 profile

Reduced value of the parameter of section classification for steel grade S235

$$\varepsilon = 0.85 \sqrt{\frac{235}{f_y}} = 0.85 \sqrt{\frac{235}{235}} = 0.85$$

The class of the web in compression

$$\frac{c}{t_w} = \frac{16.4}{0.75} = 21.87 < 33\varepsilon = 28.05 \rightarrow Class\ 1$$

The class of the flange in compression

$$\frac{c}{t_f} = \frac{9.52}{1.20} = 7.94 < 10\varepsilon = 8.5 \rightarrow Class\ 2$$

The cross section is Class 2!

Example 2.4

Calculate the cross section class for HEA 240 in steel grade S235 exposed to fire. Consider simply supported beam subjected to axial compression force of 250 kN and bending moment of 50 kNm.

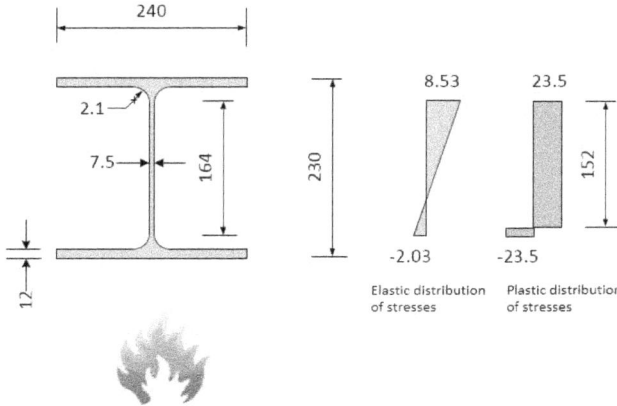

Fig. 2.5 Cross section dimensions including elastic and plastic stress diagrams for unprotected HEA-240 profile

Stress coefficients from diagrams can be calculated using EN 1993-1-1 and Designers' Guide to EN 1993-1-1 as follows. [Ref 11]

$$\psi = \frac{-2.03}{8.53} = -0.24$$

$$\alpha = \frac{1}{c}\left[\frac{h}{2} + \frac{1}{2}\frac{N_{Ed}}{t_w f_y} - (t_f + r)\right]^{[11]}$$

$$= \frac{1}{16.4}\left[\frac{23}{2} + \frac{1}{2}\frac{250 * 10}{0.75 * 235} - (1.2 + 2.1)\right] = 0.93$$

Reduced value of the parameter of section classification for steel grade S235

$$\varepsilon = 0.85\sqrt{\frac{235}{f_y}} = 0.85\sqrt{\frac{235}{235}} = 0.85$$

Web in combined bending and compression

$$\frac{c}{t_w} = 21.86 < \frac{42 * 0.85}{0.67 - 0.33 * 0.24} = 60.42 \rightarrow Class\ 3\ or\ better$$

$$\frac{c}{t_w} = 21.86 < \frac{396 * 0.85}{13 * 0.93 - 1} = 30.35 \rightarrow Class\ 1$$

The class of the flange in compression

$$\frac{c}{t_f} = 7.94 < 10\varepsilon = 8.5 \rightarrow Class\ 2$$

The cross section is Class 2!

Example 2.5

Calculate the cross section class for IPE 300 in steel grade S460 exposed to fire. Consider simply supported column with pure compression!

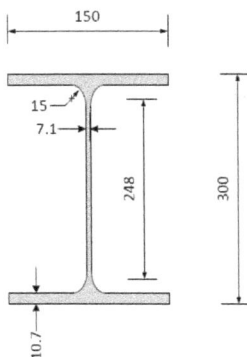

Fig. 2.6 Cross section dimensions for unprotected IPE-300 profile

Reduced value of the parameter of section classification for steel grade S460

$$\varepsilon = 0.85 \sqrt{\frac{235}{f_y}} = 0.85 \sqrt{\frac{235}{460}} = 0.61$$

The class of the web in compression

$$\frac{c}{t_w} = \frac{24.8}{0.71} = 34.93 < 42\varepsilon = 25.62 \rightarrow Class\ 4$$

The class of the flange in compression

$$\frac{c}{t_f} = \frac{5.64}{1.07} = 5.27 < 9\varepsilon = 5.49 \rightarrow Class\ 1$$

The cross section is Class 4!

3

DESIGN IN STRENGTH DOMAIN

In this Chapter

- Tension members
- Compression members
- Shear members
- Bending members
- Combined bending and shear
- Combined bending and axial force
- Critical moment for LTB
- Lateral torsional buckling
- Combined bending and axial compression
- Members with class 4 cross sections

Tension Members

Uniform temperature distribution is supposed throughout the steel cross section. We compare the design value of the tension force in fire load combination with the design resistance of a tension member at elevated temperature.

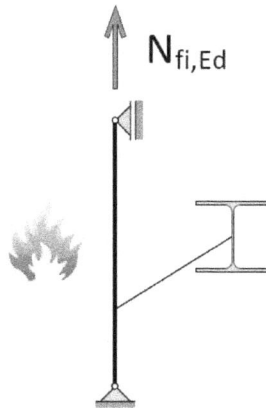

Fig. 3.1 Column subjected to pure tension

$$N_{fi,Ed} \leq= N_{fi,\vartheta,Rd} = k_{y,\vartheta} N_{Rd} \frac{\gamma_{M,0}}{\gamma_{M,fi}} = A k_{y,\vartheta} \frac{f_y}{\gamma_{M,fi}} \ldots (3.1)$$

$N_{fi,Ed}$ design value of a tension force

$N_{fi,\vartheta,Rd}$ design resistance of a tension member

A cross section area of the member

$k_{y,\vartheta}$ reduction factor for yield strength

f_y yield strength

$\gamma_{M,fi}$ material factor for fire design situation

Example 3.1

Calculate a design resistance of simply supported column constructed from HEA-240 profile of steel grade S235. Column is subjected to axial tension force of 250kN at the temperature of 650 degrees Celsius.

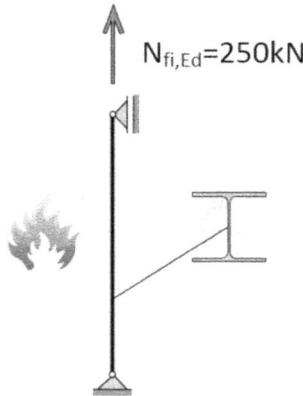

Fig. 3.2 Column subjected to tension force

Cross section and yield strength of steel are

$$A = 76.8 \ cm^2 \quad f_y = 235 \frac{N}{mm^2}$$

Reduction factor for yield strength can be calculated by interpolation from EN 1993-1-2 / Table 3.1 as follows

$$k_{y,\vartheta} = 0.47 - \frac{50 * 0.24}{100} = 0.35$$

Design resistance of column subjected to tension force is

$$N_{fi,\vartheta,Rd} = Ak_{y,\vartheta} \frac{f_y}{\gamma_{M,fi}} = 76.8 * 0.35 \frac{235}{1.0 * 10} = 631.68kN$$

Strength domain verification

$$N_{fi,Ed} = 250kN \leq= N_{fi,\vartheta,Rd} = 631.68kN$$

Compression Members
(with classes 1, 2, 3)

Uniform temperature distribution is supposed throughout the steel cross section. We compare the design value of the compression force in fire load combination with the design buckling resistance of a compression member at elevated temperature.

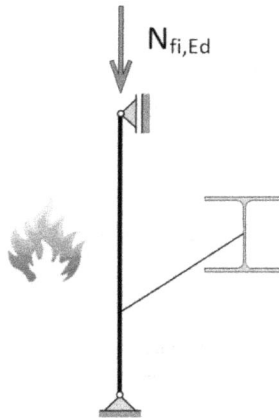

Fig. 3.3 Column subjected to pure compression

$$N_{fi,Ed} \leq= N_{b,fi,\vartheta,Rd} = k_{y,\vartheta}\chi_{fi}A \frac{f_y}{\gamma_{M,fi}} \dots (3.2)$$

$$\chi_{fi} = \frac{1}{\varphi_\vartheta + \sqrt{\varphi_\vartheta^2 - \bar{\lambda}_\vartheta^2}} \qquad \varphi_\vartheta = \frac{1}{2}\left(1 + \alpha\bar{\lambda}_\vartheta + \bar{\lambda}_\vartheta^2\right)$$

$$\alpha = 0.65 \sqrt{\frac{235}{f_y}} \qquad \bar{\lambda}_\vartheta = \bar{\lambda} \sqrt{\frac{k_{y,\vartheta}}{k_{E,\vartheta}}} \qquad \bar{\lambda} = \frac{\lambda}{\lambda_1} \qquad \lambda = \frac{l_{fi}}{i}$$

$$\lambda_1 = \pi \sqrt{\frac{E}{f_y}} = 93.9\varepsilon \quad \varepsilon = \sqrt{\frac{235}{f_y}}$$

Buckling length for fire design situation should be calculated on the same way as for the room temperature.

$N_{fi,Ed}$	design value of a compression force
$N_{b,fi,\vartheta,Rd}$	design buckling resistance of a compression member
A	cross section area of the member
$k_{y,\vartheta}$	reduction factor for yield strength
$k_{E,\vartheta}$	reduction factor for elastic modulus
χ_{fi}	reduction factor for flexural buckling
l_{fi}	buckling length
λ	member slenderness at room temperature
$\overline{\lambda}_\vartheta$	member slenderness at elevated temperature
E	elastic modulus of steel
f_y	yield strength
$\gamma_{M,fi}$	material factor for fire design situation

Example 3.2

Calculate a design resistance of simply supported column constructed from HEA-240 profile of steel grade S235 and length of 4m. Column is subjected to axial compression force of 250 kN at the temperature of 650 degrees Celsius. It is assumed that a column is part of a braced structure so buckling length is taken as system length of the column.

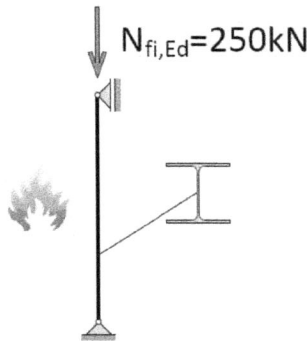

Fig. 3.4 Column subjected to compression force

Cross section, governing radius of gyration and yield strength of steel are as follows

$$A = 76.8 \ cm^2 \quad i_y = 10.05 cm \quad i_z = 6 cm \ f_y = 235 \frac{N}{mm^2}$$

Reduction factors for yield strength and Young's modulus can be calculated by interpolation from EN 1993-1-2 / Table 3.1 as follows

$$k_{y,\vartheta} = 0.47 - \frac{50 * 0.24}{100} = 0.35$$

$$k_{E,\vartheta} = 0.31 - \frac{50 * 0.18}{100} = 0.22$$

The cross section is Class 2(see example 2.3). Non-dimensional slenderness at room temperature is calculated as follows

$$\varepsilon = \sqrt{\frac{235}{235}} \quad \lambda_1 = 93.9\varepsilon = 93.9$$

$$\lambda_y = \frac{l_{fi}}{i_y} = \frac{400}{10.05} = 39.8 \quad \rightarrow \quad \bar{\lambda}_y = \frac{\lambda_y}{\lambda_1} = \frac{39.8}{93.9} = 0.42$$

$$\lambda_z = \frac{l_{fi}}{i_z} = \frac{400}{6} = 66.67 \quad \rightarrow \quad \bar{\lambda}_z = \frac{\lambda_z}{\lambda_1} = \frac{66.67}{93.9} = 0.71$$

Non-dimensional slenderness at elevated temperature is calculated as follows

$$\bar{\lambda}_{\vartheta,y} = \bar{\lambda}_y \sqrt{\frac{k_{y,\vartheta}}{k_{E,\vartheta}}} = 0.42 \sqrt{\frac{0.35}{0.22}} = 0.53$$

$$\bar{\lambda}_{\vartheta,z} = \bar{\lambda}_z \sqrt{\frac{k_{y,\vartheta}}{k_{E,\vartheta}}} = 0.71 \sqrt{\frac{0.35}{0.22}} = 0.89$$

Reduction factors for flexural buckling are calculated as follows

$$\alpha = 0.65 \sqrt{\frac{235}{235}} = 0.65$$

$$\varphi_{\vartheta,y} = \frac{1}{2}\left(1 + \alpha\bar{\lambda}_{\vartheta,y} + \bar{\lambda}_{\vartheta,y}^2\right) = \frac{1}{2}(1 + 0.65 * 0.53 + 0.53^2) = 0.81$$

$$\varphi_{\vartheta,z} = \frac{1}{2}\left(1 + \alpha\bar{\lambda}_{\vartheta,z} + \bar{\lambda}_{\vartheta,z}^2\right) = \frac{1}{2}(1 + 0.65 * 0.89 + 0.89^2) = 1.18$$

$$\chi_{fi,y} = \cfrac{1}{\varphi_{\vartheta,y} + \sqrt{\varphi_{\vartheta,y}^2 - \bar{\lambda}_{\vartheta,y}^2}} = \cfrac{1}{0.81 + \sqrt{0.81^2 - 0.53^2}} = 0.70$$

$$\chi_{fi,z} = \cfrac{1}{\varphi_{\vartheta,z} + \sqrt{\varphi_{\vartheta,z}^2 - \bar{\lambda}_{\vartheta,z}^2}} = \cfrac{1}{1.18 + \sqrt{1.18^2 - 0.89^2}} = 0.51$$

Design resistance of a compression member is calculated as follows

$$N_{b,fi,\vartheta,Rd} = k_{y,\vartheta} \chi_{fi} A \frac{f_y}{\gamma_{M,fi}} = 0.35 * 0.51 * \frac{235}{1.0 * 10} = 322.2 kN$$

Strength domain verification

$$N_{fi,Ed} = 250 kN \leq= N_{b,fi,\vartheta,Rd} = 322.2 kN$$

Shear Members
(with classes 1, 2, 3)

Uniform temperature distribution is supposed throughout the steel cross section. We compare the design value of the shear force in fire load combination with the design shear resistance of member at elevated temperature.

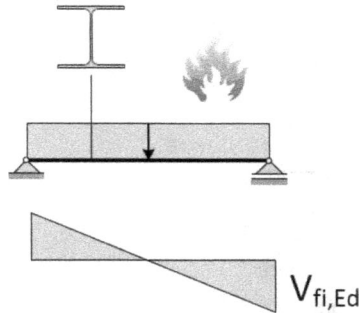

Fig. 3.5 Beam subjected to shear

$$V_{fi,Ed} \leq V_{\vartheta,Rd} = k_{y,\vartheta} V_{Rd} \frac{\gamma_{M,0}}{\gamma_{M,fi}} = \frac{k_{y,\vartheta} A_v f_y}{\sqrt{3}\gamma_{M,fi}} \dots (3.3)$$

$V_{fi,Ed}$ design value of a shear force

$V_{\vartheta,Rd}$ design shear resistance of a member

A_v shear cross section area of the member

$k_{y,\vartheta}$ reduction factor for yield strength

f_y yield strength

$\gamma_{M,fi}$ material factor for fire design situation

Shear cross section area of the member is calculated in Eurocode 3 first part (EN 1993-1-1) and we will not go into detail here as it is part of the design of steel structure at room temperature.

Example 3.3

Calculate a design resistance of simply supported beam constructed from HEA-240 profile of steel grade S235. Beam is subjected to shear force of 100 kN at the temperature of 650 degrees Celsius.

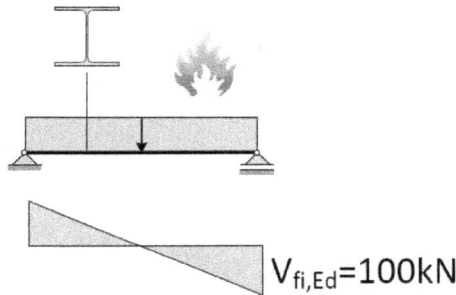

Fig. 3.6 Beam subjected to shear force

Cross sectional shear area and yield strength of steel are

$$A_v = 25.18 \; cm^2 \quad f_y = 235 \frac{N}{mm^2}$$

Reduction factor for yield strength can be calculated by interpolation from EN 1993-1-2 / Table 3.1 as follows

$$k_{y,\vartheta} = 0.47 - \frac{50 * 0.24}{100} = 0.35$$

Design shear resistance of a member is

$$V_{\vartheta,Rd} = \frac{k_{y,\vartheta} A_v f_y}{\sqrt{3}\gamma_{M,fi}} = \frac{0.35 * 25.18 * 235}{\sqrt{3} * 1.0 * 10} = 119.6 kN$$

Strength domain verification

$$V_{fi,Ed} = 100 kN \leq V_{\vartheta,Rd} = 119.6 kN$$

Bending Members (with classes 1, 2, 3)

We compare the design value of the bending moment in fire load combination with the design moment resistance of a member at elevated temperature.

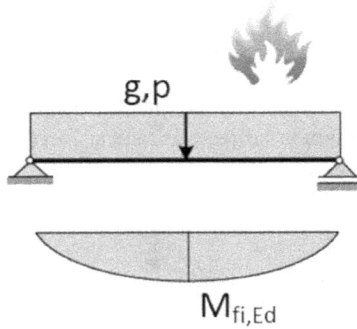

Fig. 3.7 Beam subjected to bending

$$M_{fi,Ed} \leq M_{\vartheta,Rd} = k_{y,\vartheta} \frac{\gamma_{M,0}}{\gamma_{M,fi}} \frac{M_{Rd}}{k_1 k_2} \leq M_{Rd} \dots (3.4)$$

$$M_{Rd} = \frac{W_i f_y}{\gamma_{M,0}} \qquad W_i = \begin{Bmatrix} W_{pl} & Classes\ 1,2 \\ W_{el} & Classes\ 3 \end{Bmatrix}$$

$M_{fi,Ed}$	design value of a bending moment
$M_{\vartheta,Rd}$	design moment resistance at elevated temperature
M_{Rd}	design moment resistance at room temperature
$k_{y,\vartheta}$	reduction factor for yield strength
k_1, k_2	adaptation factors
f_y	yield strength
W_i	section modulus
$\gamma_{M,0}$	material factor for room design situation
$\gamma_{M,fi}$	material factor for fire design situation

Adaptation factor for non-uniform temperature distribution along the beam k_2 and adaptation factor for non-uniform temperature distribution across the cross section k_1 are presented in tables below. Non-linear temperature distribution can be according to Eurocodes calculated only for cross sections of classes 1, 2 as there are no expressions for classes 3 or 4. You can get more detailed information about that in ECCS Eurocode Design Manual. [11]

Adaptation factor	k_1
Beam exposed on all four sides	1.00
Unprotected beam exposed on 3 sides	0.70
Protected beam exposed on 3 sides	0.85

Adaptation factor	k_2
Supports of statically indeterm. beam	0.85
All other cases	1.00

Example 3.4

Calculate design bending resistance of a simply supported beam constructed from HEA-240 profile of steel grade S235, supporting the concrete slab. Beam is subjected to bending moment of 50kNm at the temperature of 650 degrees Celzius.

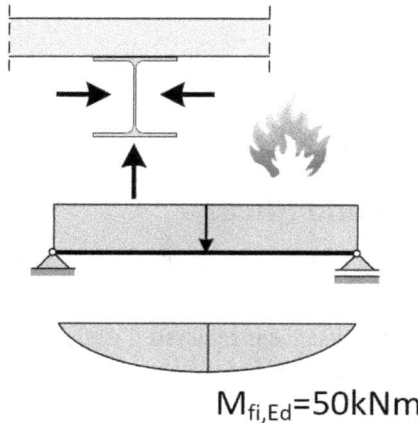

$M_{fi,Ed}=50kNm$

Fig. 3.8 Beam subjected to bending moment

The cross section is Class 2(see example 2.2).

Plastic section modulus and yield strength of steel are as follows

$$W_{pl} = 744.6cm^3 \quad f_y = 235\frac{N}{mm^2}$$

Reduction factor for yield strength can be calculated by interpolation from EN 1993-1-2 / Table 3.1 as follows

$$k_{y,\vartheta} = 0.47 - \frac{50 * 0.24}{100} = 0.35$$

Design moment resistance of a member at room temperature can be calculated as follows

$$M_{Rd} = \frac{W_i f_y}{\gamma_{M,0}} = \frac{744.6 * 235}{1.0 * 1000} = 174.98 kNm$$

As we are checking middle span section for maximum bending moment we have adaptation factor

$$k_2 = 1.0$$

As we have unprotected beam exposed on three sides we have adaptation factor

$$k_1 = 0.7$$

Design moment resistance of a member can be calculated as

$$M_{\vartheta,Rd} = k_{y,\vartheta} \frac{\gamma_{M,0}}{\gamma_{M,fi}} \frac{M_{Rd}}{k_1 k_2} = 0.35 \frac{174.98}{0.7 * 1.0} = 87.49 kNm \leq M_{Rd}$$
$$= 174.98 kNm$$

Strength domain verification

$$M_{fi,Ed} = 50 kNm \leq M_{\vartheta,Rd} = 87.49 kNm$$

Combined Bending and Shear

When shear force and bending moment act together, moment resistance is reduced due to shear.

No reduction of the moment resistance is required if design shear force in fire design combination is less than or equal of half of the shear resistance of the member as follows

$$V_{fi,Ed} \leq 0.5 V_{fi,\vartheta,Rd} \text{ ... (3.5)}$$

If the upper condition is not fulfilled, moment resistance should be reduced due to shear according to ECCS Design Manual as follows. [Ref 10]

$$M_{y,fi,Ed} \leq M_{y,V,\vartheta,Rd} \text{ ... (3.6)} \qquad A_w = h_w t_w$$

$$M_{y,V,\vartheta,Rd} = \frac{\left[W_{pl,y} - \left(\dfrac{2V_{y,fi,Ed}}{\dfrac{A_v k_{y,\vartheta} f_y}{\sqrt{3}\gamma_{M.fi}}} - 1 \right)^2 \dfrac{A_w^2}{4t_w} \right]}{k_1 k_2 \gamma_{M.fi}} k_{y,\vartheta} f_y \quad \ldots (3.7)$$

$$M_{z,fi,Ed} \leq M_{z,V,\vartheta,Rd} \quad \ldots (3.8) \qquad A_f = 2bt_f$$

$$M_{z,V,\vartheta,Rd} = \frac{\left[W_{pl,z} - \left(\dfrac{2V_{z,fi,Ed}}{\dfrac{A_v k_{y,\vartheta} f_y}{\sqrt{3}\gamma_{M.fi}}} - 1 \right)^2 \dfrac{A_f^2}{8t_f} \right]}{k_1 k_2 \gamma_{M.fi}} k_{y,\vartheta} f_y \quad \ldots (3.9)$$

Equations are valid for doubly symmetrical I cross sections of Classes 1, 2.

Example 3.5

Calculate design bending resistance and consider the effects of shear on the bending resistance for beam constructed from HEA-240 profile of steel grade S235. Beam is subjected to bending moment of 50kNm and shear force of 75kN at the temperature of 650 degrees Celsius.

The cross section is Class 2(see example 2.2).

Plastic section modulus, shear area and yield strength of steel are as follows

$$W_{pl} = 744.6 cm^3 \quad A_v = 25.18 cm^2 \quad f_y = 235 \frac{N}{mm^2}$$

Reduction factor for yield strength can be calculated by interpolation from EN 1993-1-2 / Table 3.1 as follows

$$k_{y,\vartheta} = 0.47 - \frac{50 * 0.24}{100} = 0.35$$

Design shear resistance of a member is calculated as

$$V_{fi,\vartheta,Rd} = \frac{k_{y,\vartheta} A_v f_y}{\sqrt{3}\gamma_{M,fi}} = \frac{0.35 * 25.18 * 235}{\sqrt{3} * 1.0 * 10} = 119.6 kN$$

Reduction of the moment resistance is required

$$V_{fi,Ed} = 75kN > 0.5 V_{fi,\vartheta,Rd} = 59.8 kN$$

Design bending resistance of a member reduced due to shear is

$$M_{y,V,\vartheta,Rd} = \frac{\left[W_{pl,y} - \left(\dfrac{2V_{y,fi,Ed}}{\dfrac{A_v k_{y,\vartheta} f_y}{\sqrt{3}\gamma_{M.fi}}} - 1 \right)^2 \dfrac{A_w^2}{4t_w} \right]}{k_1 k_2 \gamma_{M.fi}} k_{y,\vartheta} f_y$$

$$= \frac{\left[744.6 - \left(\dfrac{2 * 75 * 10}{\dfrac{25.18 * 0.35 * 235}{\sqrt{3} * 1.0}} - 1 \right)^2 \dfrac{15.45^2}{4 * 0.75} \right]}{1.0 * 1.0 * 1.0} \frac{0.35 * 235}{1000}$$

$$= 60.82 kNm$$

Strength domain verification

$$M_{y,fi,Ed} = 50kNm \le M_{y,V,\vartheta,Rd} = 60.82kNm$$

Combined Bending and Axial Load

When axial force and bending moment act together, moment resistance may be reduced due to axial force. According to ECCS Design manual we can use the following expression. [Ref 10]

$$M_{fi,Ed} \le M_{N,\vartheta,Rd} \ldots (3.10)$$

$$M_{N,\vartheta,Rd} = \frac{\left(1 - \dfrac{N_{fi,Ed}}{A k_{y,\vartheta} \dfrac{f_y}{\gamma_{M,fi}}} \right)}{(1 - 0.5a)} W_{pl,y} k_{y,\vartheta} \frac{f_y}{\gamma_{M,fi}} \ldots (3.11)$$

$$a = \frac{(A - 2bt_f)}{A} \le 0.5 \ldots (3.12)$$

Equations are valid for doubly symmetrical I and H cross sections with Classes 1, 2.

Example 3.6

Calculate design bending resistance and consider the effects of axial force on the bending resistance for beam constructed from HEA-240 profile of steel grade S235. Beam is subjected to bending moment of 50kNm and axial force of 250kN at the temperature of 650 degrees Celzius.

The cross section is Class 2(see example 2.4).

Plastic section modulus, cross section area and yield strength of steel are

$$W_{pl} = 744.6cm^3 \quad A = 76.84cm^2 \quad f_y = 235\frac{N}{mm^2}$$

Reduction factor for yield strength can be calculated by interpolation from EN 1993-1-2 / Table 3.1 as follows

$$k_{y,\vartheta} = 0.47 - \frac{50 * 0.24}{100} = 0.35$$

Design moment resistance of a member is calculated as

$$a = \frac{(A - 2bt_f)}{A} = \frac{(76.84 - 2 * 24 * 1.2)}{76.84} = 0.25 \le 0.5$$

$$M_{N,\vartheta,Rd} = \frac{\left(1 - \dfrac{N_{fi,Ed}}{Ak_{y,\vartheta}\dfrac{f_y}{\gamma_{M,fi}}}\right)}{(1 - 0.5a)} W_{pl,y}k_{y,\vartheta}\frac{f_y}{\gamma_{M,fi}}$$

$$= \frac{\left(1 - \dfrac{150 * 10}{76.84 * 0.35\dfrac{235}{1.0}}\right)}{(1 - 0.5 * 0.25)} 744.6 * 0.35 * \frac{235}{1000}$$

$$= 53.38kNm$$

Strength domain verification

$$M_{fi,Ed} = 42.3kNm \le M_{N,\vartheta,Rd} = 53.38kNm$$

Critical Moment for LTB

Let's learn, how we can calculate critical moment for lateral torsional buckling for uniform symmetrical cross sections. Calculation of this moment is not part of this course, we will just review this knowledge, because we need it for calculation of LTB resistance at elevated temperature.

For calculating the critical moment we can use the following formula which is not covered with Eurocodes, but is available in Designers' Guide to Eurocode 3. [Ref 11]

We will here just explain the formula shortly, more detailed information can be found in reference title.

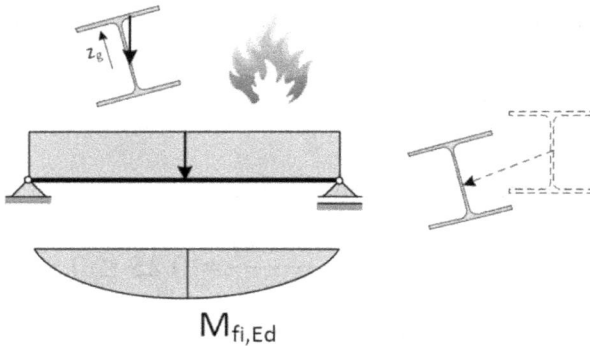

Fig. 3.9 Lateral torsional buckling of beam

$$M_{cr} = C_1 \frac{\pi^2 E I_z}{(k_z L)^2} \sqrt{\left(\frac{k_z}{k_\omega}\right)^2 \frac{I_\omega}{I_z} + \frac{(k_z L)^2 G I_t}{\pi^2 E I_z} + \left(C_2 z_g\right)^2} - C_2 z_g \quad \text{... (3.13)}$$

C_1, C_2 factor of the shape of bending moment diagram
E elastic modulus
G shear modulus
I_z second moment of area about weak axis
I_ω warping constant
I_t torsional constant
L unrestrained length of the beam
k_z effective length factor for lateral bending
k_ω effective length factor for warping restraint
z_g level of the application of load

If the load is applied at the section centroid, we can take the following simplification

$$z_g = 0$$

For normal restraint conditions at beam ends, we can take the following simplification

$$k_z = k_\omega = 1$$

Example 3.7

Calculate elastic critical moment for lateral torsional buckling for beam constructed from HEA-240 profile of steel grade S235. Beam has the length of 5m and is lateral unrestrained.

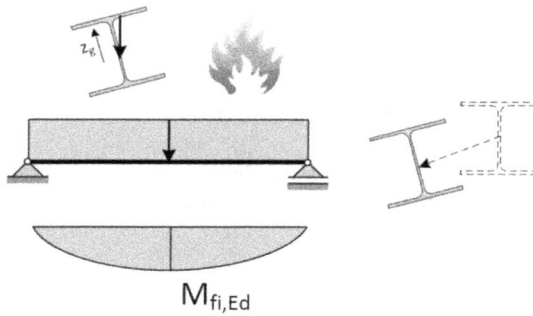

Fig. 3.10 Critical moment calculation for beam

Young's modulus, shear modulus

$$E = 21000kN/cm^2 \quad G = 8076.92kN/cm^2$$

Second moment of area, warping constant, torsion constant

$$I_z = 2769.0cm^4 \quad I_\omega = 328500.0cm^6 \quad I_t = 41.55cm^4$$

The normal conditions of restraints are assumed at both ends

$$k_z = k_\omega = 1.0$$

Equivalent uniform moment factor

$$C_1 = 1.12$$

Load is applied at the section centroid

$$z_g = 0$$

Elastic critical moment for lateral torsional buckling is

$$M_{cr} = C_1 \frac{\pi^2 E I_z}{(k_z L)^2} \sqrt{\left(\frac{k_z}{k_\omega}\right)^2 \frac{I_\omega}{I_z} + \frac{(k_z L)^2 G I_t}{\pi^2 E I_z} + \left(C_2 z_g\right)^2} - C_2 z_g$$

$$= 1.12 \frac{\pi^2 21000 * 2769}{(1.0 * 5)^2 * 10^4}$$

$$* \sqrt{\left(\frac{1.0}{1.0}\right)^2 \frac{328500}{2769 * 10^4} + \frac{(1.0 * 5)^2 8076.92 * 41.55}{\pi^2 21000 * 2769}}$$

$$= 1022.16 kNm$$

LTB of Beams with Classes 1, 2, 3

We compare the design value of the bending moment in fire load combination with the design lateral torsional buckling resistance of a member at elevated temperature.

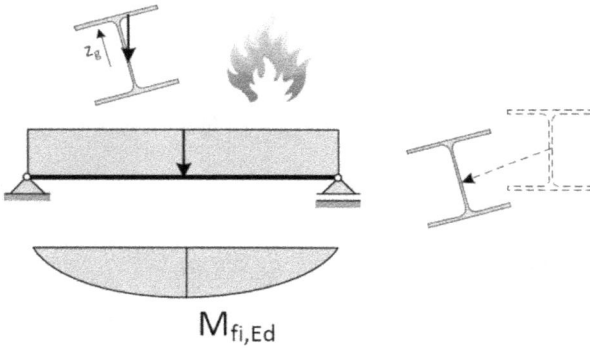

Fig. 3.11 Lateral torsional buckling of beam

$$M_{fi,Ed} \leq M_{b,\vartheta,Rd} = \chi_{LT,fi} W_y \, k_{y,\vartheta,com} \frac{f_y}{\gamma_{M,fi}} \dots (3.14)$$

$$W_y = \begin{cases} W_{pl} & Classes \ 1,2 \\ W_{el} & Classes \ 3 \end{cases}$$

$$\chi_{LT,fi} = \frac{1}{\Phi_{LT,\vartheta,com} + \sqrt{\Phi_{LT,\vartheta,com}^2 - \bar{\lambda}_{LT,\vartheta,com}^2}}$$

$$\Phi_{LT,\vartheta,com} = \frac{1}{2} \left(1 + \alpha \bar{\lambda}_{LT,\vartheta,com} + \bar{\lambda}_{LT,\vartheta,com}^2 \right)$$

$$\alpha = 0.65 \sqrt{\frac{235}{f_y}} \quad \overline{\lambda}_{LT,\vartheta,com} = \overline{\lambda}_{LT} \sqrt{\frac{k_{y,\vartheta,com}}{k_{E,\vartheta,com}}}$$

$$\overline{\lambda}_{LT} = \sqrt{\frac{W_y f_y}{M_{cr}}}$$

$M_{fi,Ed}$	design value of a bending moment
$M_{b,\vartheta,Rd}$	design LTB moment resistance at elevated temperature
M_{cr}	critical moment for LTB
$\chi_{LT,fi}$	reduction factor for LTB
W_y	section modulus
$k_{y,\vartheta,com}$	reduction factor for yield strength at max. temperature
$k_{E,\vartheta,com}$	reduction factor for Young's modulus at max. temperature
f_y	yield strength
$\gamma_{M,0}$	material factor for room design situation
$\gamma_{M,fi}$	material factor for fire design situation

Example 3.8

Calculate design lateral torsional buckling resistance for a beam constructed from HEA-240 profile of steel grade S235 at the temperature of 650 degrees Celsius. Beam has the length of 5m and is lateral unrestrained.

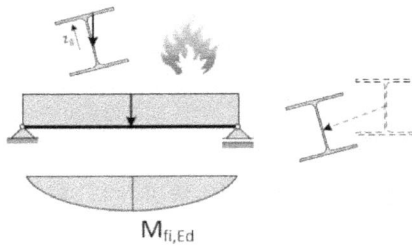

$M_{fi,Ed}$

Fig. 3.12 Lateral torsional buckling calculation of beam

The cross section is Class 2(see example 2.2).

Plastic section modulus and yield strength of steel are as follows

$$W_{pl} = 744.6cm^3 \quad f_y = 235\frac{N}{mm^2}$$

Reduction factors for yield strength and Young's modulus can be calculated by interpolation from EN 1993-1-2 / Table 3.1 as follows

$$k_{y,\vartheta} = 0.47 - \frac{50 * 0.24}{100} = 0.35$$

$$k_{E,\vartheta} = 0.31 - \frac{50 * 0.18}{100} = 0.22$$

Non-dimensional slenderness for LTB at room temperature

$$\overline{\lambda}_{LT} = \sqrt{\frac{W_y f_y}{M_{cr}}} = \sqrt{\frac{744.6 * 235}{1022.16 * 10^3}} = 0.41$$

Non-dimensional slenderness for LTB at elevated temperature

$$\overline{\lambda}_{LT,\vartheta,com} = \overline{\lambda}_{LT} \sqrt{\frac{k_{y,\vartheta,com}}{k_{E,\vartheta,com}}} = 0.41 \sqrt{\frac{0.35}{0.22}} = 0.52$$

LTB reduction factors at elevated temperature

$$\alpha = 0.65 \sqrt{\frac{235}{235}} = 0.65$$

$$\Phi_{LT,\vartheta,com} = \frac{1}{2}\left(1 + \alpha \overline{\lambda}_{LT,\vartheta,com} + \overline{\lambda}^2_{LT,\vartheta,com}\right)$$
$$= \frac{1}{2}(1 + 0.65 * 0.52 + 0.52^2) = 0.80$$

LTB reduction factor at elevated temperature

$$\chi_{LT,fi} = \frac{1}{\Phi_{LT,\vartheta,com} + \sqrt{\Phi^2_{LT,\vartheta,com} - \overline{\lambda}^2_{LT,\vartheta,com}}} = \frac{1}{0.8 + \sqrt{0.8^2 - 0.52^2}}$$
$$= 0.71$$

Design LTB moment resistance of a member at elevated temperature can be calculated as follows

$$M_{b,\vartheta,Rd} = \chi_{LT,fi} W_y k_{y,\vartheta,com} \frac{f_y}{\gamma_{M,fi}} = 0.71 * 744.6 * 0.35 * \frac{235}{1.0} * 10^{-3}$$
$$= 43.48 kNm$$

Combined Bending and Axial Compression

The following two equations should be fulfilled for steel elements to ensure required design resistance in fire design situation for elements with classes 1, 2, 3.

$$\frac{N_{fi,Ed}}{\chi_{min,fi}Ak_{y,\vartheta}\dfrac{f_y}{\gamma_{M,fi}}} + \frac{k_y M_{y,fi,Ed}}{W_y k_{y,\vartheta}\dfrac{f_y}{\gamma_{M,fi}}} + \frac{k_z M_{z,fi,Ed}}{W_z k_{y,\vartheta}\dfrac{f_y}{\gamma_{M,fi}}} \leq 1 \dots (3.15)$$

$$\frac{N_{fi,Ed}}{\chi_{z,fi}Ak_{y,\vartheta}\dfrac{f_y}{\gamma_{M,fi}}} + \frac{k_{LT} M_{y,fi,Ed}}{\chi_{LT,fi}W_y k_{y,\vartheta}\dfrac{f_y}{\gamma_{M,fi}}} + \frac{k_z M_{z,fi,Ed}}{W_z k_{y,\vartheta}\dfrac{f_y}{\gamma_{M,fi}}} \leq 1 \dots (3.16)$$

$$W_{y,z} = \begin{Bmatrix} W_{pl} \; Classes \; 1,2 \\ W_{el} \; Classes \; 3 \end{Bmatrix}$$

$$\chi_{min,fi} = min\{\chi_{y,fi}, \chi_{z,fi}\}$$

$$k_{LT} = 1 - \frac{\mu_{LT} N_{fi,Ed}}{\chi_{z,fi}Ak_{y,\vartheta}\dfrac{f_y}{\gamma_{M,fi}}} \leq 1$$

$$k_y = 1 - \frac{\mu_y N_{fi,Ed}}{\chi_{y,fi}Ak_{y,\vartheta}\dfrac{f_y}{\gamma_{M,fi}}} \leq 3$$

$$k_z = 1 - \frac{\mu_z N_{fi,Ed}}{\chi_{z,fi}Ak_{y,\vartheta}\dfrac{f_y}{\gamma_{M,fi}}} \leq 3$$

$$\mu_{LT} = 0.15\bar{\lambda}_{z,\vartheta}\beta_{M,LT} - 0.15 \leq 0.9$$

$$\mu_y = \left(1.2\beta_{M,y} - 3\right)\bar{\lambda}_{y,\vartheta} + 0.44\beta_{M,y} - 0.29 \leq 0.8$$

$$\mu_z = \left(2\beta_{M,z} - 5\right)\bar{\lambda}_{z,\vartheta} + 0.44\beta_{M,z} - 0.29 \leq 0.8$$

$$\bar{\lambda}_{z,\vartheta} \leq 1.1$$

β_M uniform moment factors

For uniform moment factors check EN 1993-1-2 / 4.2.3.5 Figure 4.2.

Example 3.9

Calculate design resistance for a beam constructed from HEA-240 profile of steel grade S235 at the temperature of 650 degrees Celsius. Beam has the length of 5m, is lateraly unrestrained and subjected to axial compression force of 250 kN and bending moment of 50 kNm.

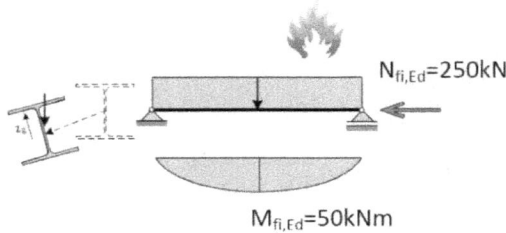

Fig. 3.13 Beam subjected to bending and compression

Cross section, governing radius of gyration and yield strength of steel are

$$A = 76.8 \, cm^2 \quad W_{pl,y} = 744.6 \, cm^3 \quad i_y = 10.05 cm$$

$$i_z = 6cm \quad f_y = 235 \frac{N}{mm^2}$$

The cross section is Class 2(see example 2.4).

Reduction factors for yield strength and Young's modulus can be calculated by interpolation from EN 1993-1-2 / Table 3.1 as follows

$$k_{y,\vartheta} = 0.47 - \frac{50 * 0.24}{100} = 0.35$$

$$k_{E,\vartheta} = 0.31 - \frac{50 * 0.18}{100} = 0.22$$

Non-dimensional slenderness at room temperature is calculated as follows

$$\varepsilon = \sqrt{\frac{235}{235}} \quad \lambda_1 = 93.9\varepsilon = 93.9$$

$$\lambda_y = \frac{l_{fi}}{i_y} = \frac{500}{10.05} = 49.75 \quad \rightarrow \quad \bar{\lambda}_y = \frac{\lambda_y}{\lambda_1} = \frac{49.75}{93.9} = 0.53$$

$$\lambda_z = \frac{l_{fi}}{i_z} = \frac{500}{6} = 83.33 \quad \rightarrow \quad \bar{\lambda}_z = \frac{\lambda_z}{\lambda_1} = \frac{83.33}{93.9} = 0.89$$

Non-dimensional slenderness at elevated temperature is calculated as follows

$$\bar{\lambda}_{\vartheta,y} = \bar{\lambda}_y \sqrt{\frac{k_{y,\vartheta}}{k_{E,\vartheta}}} = 0.53 \sqrt{\frac{0.35}{0.22}} = 0.67$$

$$\bar{\lambda}_{\vartheta,z} = \bar{\lambda}_z \sqrt{\frac{k_{y,\vartheta}}{k_{E,\vartheta}}} = 0.89 \sqrt{\frac{0.35}{0.22}} = 1.12$$

Reduction factors for flexural buckling

$$\alpha = 0.65 \sqrt{\frac{235}{235}} = 0.65$$

$$\varphi_{\vartheta,y} = \frac{1}{2}\left(1 + \alpha \bar{\lambda}_{\vartheta,y} + \bar{\lambda}_{\vartheta,y}^2\right) = \frac{1}{2}(1 + 0.65 * 0.67 + 0.67^2) = 0.94$$

$$\varphi_{\vartheta,z} = \frac{1}{2}\left(1 + \alpha \bar{\lambda}_{\vartheta,z} + \bar{\lambda}_{\vartheta,z}^2\right) = \frac{1}{2}(1 + 0.65 * 1.12 + 1.12^2) = 1.49$$

$$\chi_{fi,y} = \frac{1}{\varphi_{\vartheta,y} + \sqrt{\varphi_{\vartheta,y}^2 - \bar{\lambda}_{\vartheta,y}^2}} = \frac{1}{0.94 + \sqrt{0.94^2 - 0.67^2}} = 0.62$$

$$\chi_{fi,z} = \frac{1}{\varphi_{\vartheta,z} + \sqrt{\varphi_{\vartheta,z}^2 - \bar{\lambda}_{\vartheta,z}^2}} = \frac{1}{1.49 + \sqrt{1.49^2 - 1.12^2}} = 0.40$$

For uniform moment factors check EN 1993-1-2 / 4.2.3.5 / Figure 4.2

$$\beta_{M,y} = \beta_{M,LT} = 1.3$$

Elastic critical moment for LTB(see example 4.7)

$$M_{cr} = 1022.16 \text{kNm}$$

Non-dimensional slenderness for LTB at room temperature

$$\bar{\lambda}_{LT} = \sqrt{\frac{W_y f_y}{M_{cr}}} = \sqrt{\frac{744.6 * 235}{1022.16 * 10^3}} = 0.41$$

Non-dimensional slenderness for LTB at elevated temperature

$$\bar{\lambda}_{LT,\vartheta,com} = \bar{\lambda}_{LT}\sqrt{\frac{k_{y,\vartheta,com}}{k_{E,\vartheta,com}}} = 0.41\sqrt{\frac{0.35}{0.22}} = 0.52$$

LTB reduction factors at elevated temperature

$$\alpha = 0.65\sqrt{\frac{235}{235}} = 0.65$$

$$\Phi_{LT,\vartheta,com} = \frac{1}{2}\left(1 + \alpha\bar{\lambda}_{LT,\vartheta,com} + \bar{\lambda}^2_{LT,\vartheta,com}\right)$$

$$= \frac{1}{2}(1 + 0.65 * 0.52 + 0.52^2) = 0.80$$

$$\chi_{LT,fi} = \frac{1}{\Phi_{LT,\vartheta,com} + \sqrt{\Phi^2_{LT,\vartheta,com} - \bar{\lambda}^2_{LT,\vartheta,com}}} = \frac{1}{0.8 + \sqrt{0.8^2 - 0.52^2}}$$

$$= 0.71$$

Reduction factors

$$\mu_y = \left(1.2\beta_{M,y} - 3\right)\bar{\lambda}_{y,\vartheta} + 0.44\beta_{M,y} - 0.29$$
$$= (1.2 * 1.3 - 3) * 0.67 + 0.44 * 1.3 - 0.29 = -0.68$$
$$\leq 0.8$$

$$\mu_{LT} = 0.15 * 1.12 * 1.3 - 0.15 = 0.07 \leq 0.9$$

$$k_{LT} = 1 - \frac{0.07 * 250 * 10}{0.4 * 76.8 * 0.35 * \frac{235}{1.0}} = 0.93 \leq 1$$

$$k_y = 1 - \frac{\mu_y N_{fi,Ed}}{\chi_{y,fi}Ak_{y,\vartheta}\frac{f_y}{\gamma_{M,fi}}} = 1 + \frac{0.68 * 250 * 10}{0.62 * 76.8 * 0.35\frac{235}{1.0}} = 1.43 \leq 3$$

Interaction equations can be calculated as

$$\frac{N_{fi,Ed}}{\chi_{min,fi}Ak_{y,\vartheta}\dfrac{f_y}{\gamma_{M,fi}}} + \frac{k_y M_{y,fi,Ed}}{W_y k_{y,\vartheta}\dfrac{f_y}{\gamma_{M,fi}}}$$

$$= \frac{250 * 10}{0.4 * 76.8 * 0.35 * \dfrac{235}{1.0}} + \frac{1.43 * 50 * 10^3}{744.6 * 0.35 * \dfrac{235}{1.0}} = 2.15$$

$$> 1 \rightarrow Not\ satisfactory!$$

$$\frac{N_{fi,Ed}}{\chi_{z,fi}Ak_{y,\vartheta}\dfrac{f_y}{\gamma_{M,fi}}} + \frac{k_{LT} M_{y,fi,Ed}}{\chi_{LT,fi}W_y k_{y,\vartheta}\dfrac{f_y}{\gamma_{M,fi}}}$$

$$= \frac{250 * 10}{0.4 * 76.8 * 0.35 * \dfrac{235}{1.0}} + \frac{0.93 * 50 * 10^3}{0.71 * 744.6 * 0.35 * \dfrac{235}{1.0}}$$

$$= 2.05 > 1 \quad \rightarrow Not\ satisfactory!$$

Members with Class 4 Cross Sections

EN 1993-1-2 alows that fire resistance is achieved as long as the temperature in the cross section does not exceed 350 degrees Celsius – a different value can be set in National Annex. This way is very convenient for a designer, but offers very conservative results as it leads to strong steel profile or rich fire protection.

Anyway Eurocodes allow that verification is done, where we can use the equations for room temperature and replace cross section area with effective area and section moduluses by the effective section moduluses in accordance with EN 1993-3 and EN 1993-1-5.

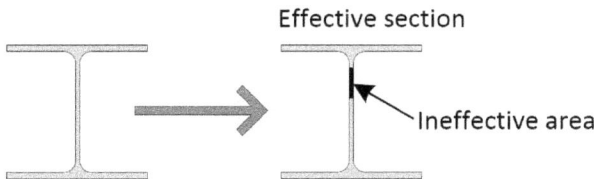

Fig. 3.14 Effective cross section for CS of class 4

$\vartheta_a\,[\degree C]$	$k_{p0.2,\vartheta}$
20	1.000
100	1.000
200	0.890
300	0.780
400	0.650
500	0.530
600	0.300
700	0.130
800	0.070
900	0.050
1000	0.030
1100	0.020
1200	0.000

If the verification is done, we need to reduce the yield strength at elevated temperature from that at room temperature using the reduction factors presented in the upper table

$$f_{y,\vartheta} = k_{p0.2,\vartheta}\, f_y \;...\,(3.17)$$

Example 3.10

Calculate design resistance for a beam constructed from IPE-300 profile of steel grade S460 at the temperature of 650 degrees Celsius. Beam has the length of 5m and is subjected to axial compression force of 50 kN.

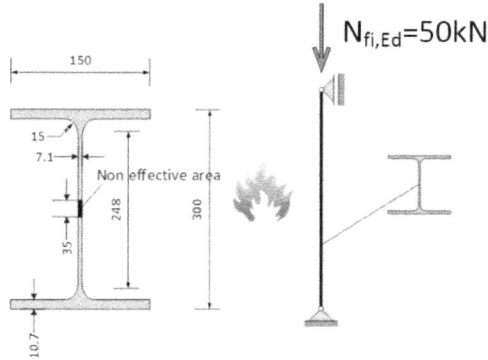

Fig. 3.15 Column of CS class 4 subjected to pure compression

Cross section area and moments of inertia are

$$A = 53.8 \ cm^2 \quad I_y = 8356 cm^4 \quad I_z = 603.8 cm^4$$

Yield strength of steel is

$$f_y = 460 \frac{N}{mm^2}$$

The cross section is Class 4 (see example 2.5).

Reduced value of the parameter of section classification for steel grade S460

$$\varepsilon = \sqrt{\frac{235}{f_y}} = \sqrt{\frac{235}{460}} = 0.71$$

For the calculation of the effective cross section we need first to evaluate the non effective part of it from EN 1993-1-5 Table 4.1 as follows.

Straight part of the web, buckling factor and stress ratio

$$\bar{b} = 24.8 cm \quad t = 0.71 cm \quad k_\sigma = 4 \quad \psi = 1$$

Normalized web slenderness

$$\bar{\lambda}_p = \frac{\bar{b}/t}{28.4\epsilon\sqrt{k_\sigma}} = \frac{24.8/0.71}{28.4 * 0.71\sqrt{4}} = 0.87$$

$$\bar{\lambda}_p = 0.87 > 0.5 + \sqrt{0.085 - 0.055\psi} = 0.67$$

Reduction factor for the effective width of the web is thus according to normalized web slenderness calculated as

$$\rho = \frac{\bar{\lambda}_p - 0.055(3 + \psi)}{\bar{\lambda}_p^{\,2}} = \frac{0.87 - 0.055(3 + 1)}{0.87^2} = 0.86$$

Effective width of the web

$$b_{eff} = \rho\bar{b} = 0.86 * 24.8 = 21.3 cm$$

Non effective width of the web

$$b_n = c - b_{eff} = 24.8 - 21.3 = 3.5 cm$$

The effective cross section area

$$A_{eff} = A - b_n t_w = 53.8 - 0.71 * 3.5 = 51.3 \ cm^2$$

The effective moments of innertia

$$I_{y,eff} = I_y - t_w b_n^3/12 = 8356 - \frac{0.71 * 3.5^3}{12} = 8353.46 \ cm^4$$

$$I_{z,eff} = I_z - t_w^3 b_n/12 = 603.8 - \frac{0.71^3 * 3.5}{12} = 603.7 \ cm^4$$

The effective radiuses of gyration

$$i_{y,eff} = \sqrt{\frac{I_{y,eff}}{A_{eff}}} = 12.76 cm \quad i_{z,eff} = \sqrt{\frac{I_{z,eff}}{A_{eff}}} = 3.43 cm$$

Reduction factors for yield strength and Young's modulus can be calculated by interpolation from EN 1993-1-2 / Table 3.1, E.1 as follows

$$k_{p0.2,\vartheta} = 0.47 - \frac{50 * 0.17}{100} = 0.21$$

$$k_{E,\vartheta} = 0.31 - \frac{50 * 0.18}{100} = 0.22$$

It is assumed that column is part of a braced structure so buckling length is taken as system length of the column. Non-dimensional slenderness at room temperature is calculated as follows

$$\lambda_1 = 93.9\varepsilon = 66.7$$

$$\lambda_y = \frac{l_{fi}}{i_y} = \frac{500}{12.76} = 39.18 \quad \rightarrow \quad \bar{\lambda}_y = \frac{\lambda_y}{\lambda_1} = \frac{39.18}{66.7} = 0.59$$

$$\lambda_z = \frac{l_{fi}}{i_z} = \frac{500}{3.43} = 145.77 \quad \rightarrow \quad \bar{\lambda}_z = \frac{\lambda_z}{\lambda_1} = \frac{145.77}{66.7} = 2.18$$

Non-dimensional slenderness at elevated temperature is calculated as follows

$$\bar{\lambda}_{\vartheta,y} = \bar{\lambda}_y \sqrt{\frac{k_{p0.2,\vartheta}}{k_{E,\vartheta}}} = 0.59 \sqrt{\frac{0.21}{0.22}} = 0.58$$

$$\bar{\lambda}_{\vartheta,z} = \bar{\lambda}_z \sqrt{\frac{k_{p0.2,\vartheta}}{k_{E,\vartheta}}} = 2.18 \sqrt{\frac{0.21}{0.22}} = 2.13$$

Reduction factors for flexural buckling are calculated as follows

$$\alpha = 0.65 \sqrt{\frac{235}{460}} = 0.46$$

$$\varphi_{\vartheta,y} = \frac{1}{2}\left(1 + \alpha\bar{\lambda}_{\vartheta,y} + \bar{\lambda}_{\vartheta,y}^2\right) = \frac{1}{2}(1 + 0.46 * 0.58 + 0.58^2) = 0.80$$

$$\varphi_{\vartheta,z} = \frac{1}{2}\left(1 + \alpha\bar{\lambda}_{\vartheta,z} + \bar{\lambda}_{\vartheta,z}^2\right) = \frac{1}{2}(1 + 0.46 * 2.13 + 2.13^2) = 3.25$$

$$\chi_{fi,y} = \frac{1}{\varphi_{\vartheta,y} + \sqrt{\varphi_{\vartheta,y}^2 - \bar{\lambda}_{\vartheta,y}^2}} = \frac{1}{0.80 + \sqrt{0.80^2 - 0.58^2}} = 0.74$$

$$\chi_{fi,z} = \frac{1}{\varphi_{\vartheta,z} + \sqrt{\varphi_{\vartheta,z}^2 - \bar{\lambda}_{\vartheta,z}^2}} = \frac{1}{3.25 + \sqrt{3.25^2 - 2.13^2}} = 0.17$$

Design resistance of a compression member is calculated as follows

$$N_{b,fi,\vartheta,Rd} = k_{p0.2,\vartheta}\chi_{fi,min}A_{Eff}\frac{f_y}{\gamma_{M,fi}} = 0.21 * 0.17 * \frac{460}{1.0 * 10} = 84.2 kN$$

Strength domain verification

$$N_{fi,Ed} = 50kN \leq= N_{b,fi,\vartheta,Rd} = 84.2kN$$

4

DESIGN IN TEMPERATURE DOMAIN

In this Chapter

- Critical temperature calculation
- Iterative procedure

Critical Temperature Calculation

Let's learn first, what is critical temperature? It is a temperature till which structural element maintains its capacity. The collapse occurs when the design temperature of a structural element euqals it's critical temperature.

Uniform temperature distribution across the cross section is assumed, what corresponds with nominal temperature-time curves of fully developed fire in fire compartment.

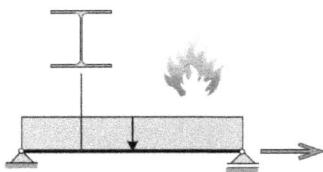

Fig. 4.1 Beam subjected to pure tension

We are thus checking if design value of material temperature is less or equal than design value of critical temperature of structural member.

$$\vartheta_{d,t} \leq \vartheta_{d,cr} \dots (4.1)$$

$$\vartheta_{d,cr} = 39.19 \, ln \left[\frac{1}{0.9674 \mu_0^{3.833}} - 1 \right] + 482 \dots (4.2)$$

$$\mu_0 = \frac{E_{fi,d}}{R_{fi,d,0}} \geq 0.013 \dots (4.3)$$

$\vartheta_{d,t}$ design value of material temperature

$\vartheta_{d,cr}$ design value of critical temperature

μ_0 degree of utilization

$E_{fi,d}$ effects of actions at fire load combination

$R_{fi,d,0}$ fire resistance at room temperature

Equation for calculation of critical temperature is introduced in EN 1993-1-2 and can be directly applied just for tension and bending elements. In all other cases, many parameters are temperature dependent and iteration procedure should be used (flexural buckling, lateral torsional buckling, etc.).

Not knowing the critical temperature leads to conservative design, especially for lower degrees of utilization factors.

Example 4.1

Calculate the critical temperature of simply supported column constructed from HEA-240 profile of steel grade S235. Column is subjected to axial tension force of 250 kN.

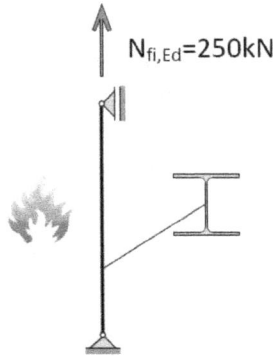

Fig. 4.2 Beam subjected to tension force

Cross section area and yield strength of steel

$$A = 76.8 \ cm^2 \quad f_y = 235 N/mm^2$$

Design resistance of a column at room temperature

$$N_{fi,d,0} = A\frac{f_y}{\gamma_{M,fi}} = 76.8\frac{235}{1.0*10} = 1804.8 kN$$

Degree of utilization

$$\mu_0 = \frac{N_{fi,d}}{R_{fi,d,0}} = \frac{250}{1804.8} = 0.14 \geq 0.013$$

Critical temperature is calculated as

$$\vartheta_{d,cr} = 39.19 \, ln\left[\frac{1}{0.9674\mu_0^{3.833}} - 1\right] + 482$$

$$= 39.19 \, ln\left[\frac{1}{0.9674 * 0.14^{3.833}} - 1\right] + 482 = 778.6°C$$

Iterative Procedure

When calculation parameters are directly dependent on temperature, equation for critical temperature cannot be applied directly and iteration procedure is required. This procedure is relevant in the following cases.

- Combined bending and compression
- Flexural buckling
- Lateral torsional buckling

Degree of utilization can be calculated according to ECCS Eurocode Design Manual, from the expressions displayed below. We will here just use these equations, more complete information and explanation can be found in reference title. [Ref 10]

$$\mu_0 = \frac{N_{fi,Ed}}{\chi_{min,fi}A\dfrac{f_y}{\gamma_{M,fi}}} + \frac{k_y M_{y,fi,Ed}}{W_y \dfrac{f_y}{\gamma_{M,fi}}} + \frac{k_z M_{z,fi,Ed}}{W_z \dfrac{f_y}{\gamma_{M,fi}}} \dots (4.4)$$

$$\mu_0 = \frac{N_{fi,Ed}}{\chi_{z,fi}A\dfrac{f_y}{\gamma_{M,fi}}} + \frac{k_{LT} M_{y,fi,Ed}}{\chi_{LT,fi}W_y \dfrac{f_y}{\gamma_{M,fi}}} + \frac{k_z M_{z,fi,Ed}}{W_z \dfrac{f_y}{\gamma_{M,fi}}} \dots (4.5)$$

The iterative procedure is used for critical temperature as shown below.

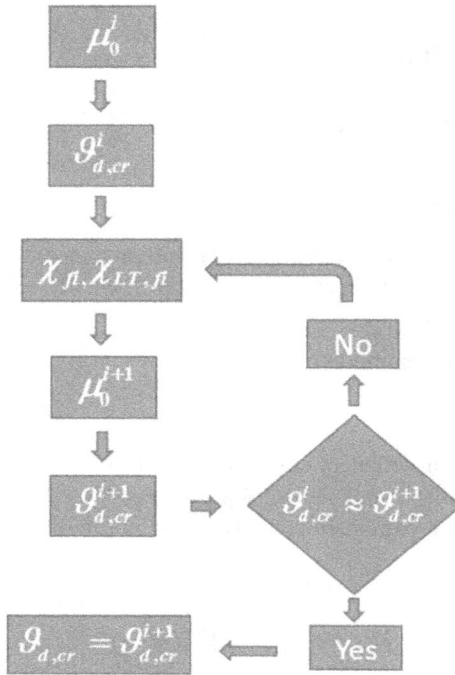

Fig. 4.3 Design chart for critical temperature calculation

Example 4.2

Calculate the critical temperature of a beam constructed from HEA-240 profile of steel grade S235. The beam has the length of 5m, is lateral restrained and subjected to axial compression force of 250 kN and bending moment of 50 kNm.

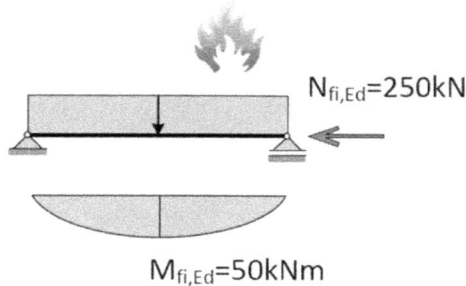

$N_{fi,Ed}=250kN$

$M_{fi,Ed}=50kNm$

Fig. 4.4 Beam subjected to combined bending and compression

Cross sectional area, plastic section modulus, governing radius of gyration and yield strength of steel are

$$A = 76.8 \ cm^2 \quad W_{pl,y} = 744.6 \ cm^3 \quad i_y = 10.05 cm$$

$$i_z = 6cm \quad f_y = 235 \frac{N}{mm^2}$$

The cross section is Class 2(see example 2.4).

It is assumed that beam is part of a braced structure, so buckling length is taken as system length of the column.

Non-dimensional slenderness at room temperature is calculated as follows

$$\varepsilon = \sqrt{\frac{235}{235}} = 1 \quad \lambda_1 = 93.9\varepsilon = 93.9$$

$$\lambda_y = \frac{l_{fi}}{i_y} = \frac{500}{10.05} = 49.75 \quad \rightarrow \quad \bar{\lambda}_y = \frac{\lambda_y}{\lambda_1} = \frac{49.75}{93.9} = 0.53$$

$$\lambda_z = \frac{l_{fi}}{i_z} = \frac{500}{6} = 83.33 \quad \rightarrow \quad \bar{\lambda}_z = \frac{\lambda_z}{\lambda_1} = \frac{83.33}{93.9} = 0.89$$

Non-dimensional slenderness at elevated temperature of first iteration, which is equal to room temperature is calculated as follows

$$\bar{\lambda}_{\vartheta=20°C,y} = \bar{\lambda}_y = 0.53 \quad \bar{\lambda}_{\vartheta=20°C,z} = \bar{\lambda}_z = 0.89$$

Reduction factors for flexural buckling

$$\alpha = 0.65\sqrt{\frac{235}{235}} = 0.65$$

$$\varphi_{\vartheta,y} = \frac{1}{2}\left(1 + \alpha\bar{\lambda}_{\vartheta,y} + \bar{\lambda}_{\vartheta,y}^2\right) = \frac{1}{2}(1 + 0.65 * 0.53 + 0.53^2) = 0.81$$

$$\varphi_{\vartheta,z} = \frac{1}{2}\left(1 + \alpha\bar{\lambda}_{\vartheta,z} + \bar{\lambda}_{\vartheta,z}^2\right) = \frac{1}{2}(1 + 0.65 * 0.89 + 0.89^2) = 1.18$$

$$\chi_{fi,y} = \frac{1}{\varphi_{\vartheta,y} + \sqrt{\varphi_{\vartheta,y}^2 - \bar{\lambda}_{\vartheta,y}^2}} = \frac{1}{0.81 + \sqrt{0.81^2 - 0.53^2}} = 0.70$$

$$\chi_{fi,z} = \cfrac{1}{\varphi_{\vartheta,z} + \sqrt{\varphi_{\vartheta,z}^2 - \bar{\lambda}_{\vartheta,z}^2}} = \cfrac{1}{1.18 + \sqrt{1.18^2 - 0.89^2}} = 0.51$$

For uniform moment factors check EN 1993-1-2 (4.2.3.5) Figure 4.2

$$\beta_{M,y} = 1.3$$

Reduction factors

$$\begin{aligned}
\mu_y &= (1.2\beta_{M,y} - 3)\bar{\lambda}_{y,\vartheta} + 0.44\beta_{M,y} - 0.29 \\
&= (1.2 * 1.3 - 3) * 0.53 + 0.44 * 1.3 - 0.29 = -0.48 \\
&\leq 0.8
\end{aligned}$$

$$k_y = 1 - \cfrac{\mu_y N_{fi,Ed}}{\chi_{y,fi} A k_{y,\vartheta} \cfrac{f_y}{\gamma_{M,fi}}} = 1 + \cfrac{0.48 * 250 * 10}{0.70 * 76.8 * 1.0\cfrac{235}{1.0}} = 1.09 \leq 3$$

Degree of utilization in first iteration can be calculated as

$$\begin{aligned}
\mu_0 &= \cfrac{N_{fi,Ed}}{\chi_{min,fi} A \cfrac{f_y}{\gamma_{M,fi}}} + \cfrac{k_y M_{y,fi,Ed}}{W_y \cfrac{f_y}{\gamma_{M,fi}}} \\
&= \cfrac{250 * 10}{0.51 * 76.8 * \cfrac{235}{1.0}} + \cfrac{1.09 * 50 * 10^3}{744.6 * \cfrac{235}{1.0}} = 0.585
\end{aligned}$$

The critical temperature in first iteration can be calculated by the following expression

$$\begin{aligned}
\vartheta_{d,cr} &= 39.19 \, ln\left[\cfrac{1}{0.9674\mu_0^{3.833}} - 1\right] + 482 \\
&= 39.19 \, ln\left[\cfrac{1}{0.9674 * 0.585^{3.833}} - 1\right] + 482 \\
&= 558.65°C
\end{aligned}$$

We are using the value of the critical temperature of 558.65 degrees Celsius for the calculation in the next iteration and continue with iterations until convergence is found.

The other iterations of calculations are presented in the table below.

ϑ^i_{cr} [°C]	$k_{y,\vartheta}$	$k_{E,\vartheta}$	μ_y	k_y	μ_o	ϑ^{i+1}_{cr} [°C]
558.65	0.598	0.194	0.648	1.221	0.667	535.01
535.01	0.671	0.498	0.653	1.190	0.653	538.99
538.99	0.659	0.487	0.652	1.195	0.655	538.40
538.40	0.661	0.489	0.652	1.194	0.655	538.49

The critical temperature is calculated after 4 iterations as 538.49 degrees Celsius.

5

FIRE PROTECTION

In this Chapter

• Fire protection systems

• Moisture in the insulation

Fire Protection Systems

What are fire protection systems and when they are required?

When the temperature of the steel structural element exceeds its critical temperature, the steel element cannot provide any more required fire safety by itself. In this case we need to protect the element so, that together with the protection insulation material it resists actual temperature of the fire compartment. It means that steel structural element is not subjected to a temperature higher as critical as protection insulation material is used for this purpose.

We have the following fire protection systems:

- Boards
- Sprays
- Intumescent coatings
- Concrete encasement

Boards

Boards are used for fire protection of beams and columns as they offer boxed appearance and are suitable for decorative finishes. Their advantage is that their application is a dry trade and as such do not have significant impact on other site operations.

Fig. 5.1 Steel elements protected with boards

Sprays

Unlike boards can be prays used to cover complex shapes of structural elements. But are not sutable for aesthtetic purposes. Their application is unlike boards a wet trade and as such have significant impact on other site operations.

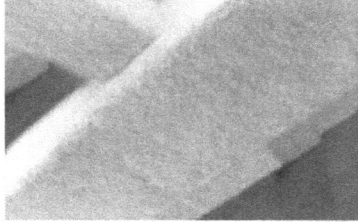

Fig. 5.2 Steel elements protected with sprays

Intumescent coatings

Intumescent coatings provide insulation unlike boards and sprays as a result of a complex chemical reaction which creates an expanded layer of low conductivity char around structural element. That way they protect structural element from high temperature of fire compartment.

Fig. 5.3 Steel cross section protected with intumescent coating

Concrete encasement

Concrete encasement is for sure one of the oldest fire protection systems for structural steelwork, whose use is dramatically reduced in the last decades. Its main advandage is durability and disadvantage is that it requires larger thickness comparing to other systems.

Fig. 5.4 Steel elements protected with concrete encasement

Thickness of insulation systems can be provided from the manufacturer on the basis of the critical temperature.

Moisture in the Insulation

As thermal conductivity of insulation material is defined for dry conditions, we should allow the time delay in steel temperature for moist insulation. According to ECCS (1983) we can use the following expression [Ref 9]

$$t_v = \frac{p\rho_p d_p^2}{5\lambda_p} \quad [min] \dots (5.1)$$

p	moisture of the insulation
ρ_p	unit mass of the insulation
λ_p	thermal conductivity of insulation
d_p	insulation thickness

Example 5.1

Calculate time delay due to moisture in insulation material of fibre-silicate boards, when the moisture content in the insulation is 15% and the insulation thickness is 2cm.

$\lambda_p = 0.15 W/mK$ thermal conductivity of insulation

$\rho_p = 600 kg/m^3$ unit mass of the insulation

$$t_v = \frac{p\rho_p d_p^2}{5\lambda_p} = \frac{15 * 600 * 0.02^2}{5 * 0.15} = 4.8 \ min$$

6

WORKED EXAMPLE

In this Chapter

- Structure definition
- Fire resistance of primary beams
- Fire resistance of secondary beams
- Fire resistance of columns
- Analysis of results and conclusions

Structure Definition

We have 2-storey office building with continuous composite 2-span slab. Three main structural elements will be analyzed for standard fire resistance of R30 where critical temperature and required fire protection will be calculated. At the end we will analyze the results and provide necessary conclusions.

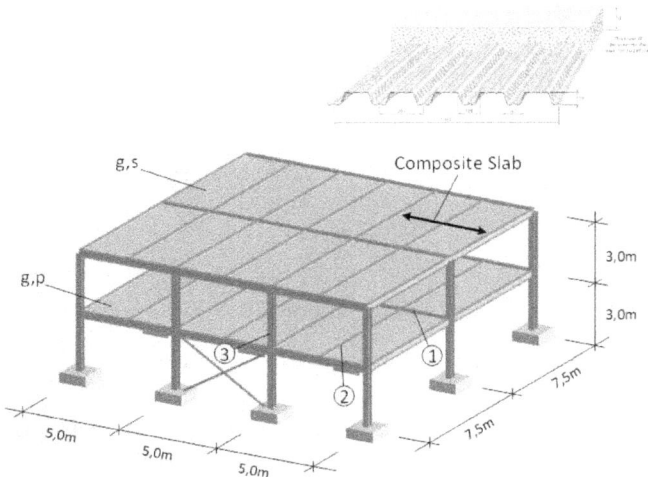

Fig. 6.1 Structure definition

Each storey has 2-span continuous composite slab with depth of 12cm of type COFRAPLUS 60. Primary beams 1 are made from HR steel profiles IPE-360 in steel quality S275. Secondary beams 2 are made from HR steel profiles IPE-330 in steel quality S275. Columns 3 are made from HR steel profiles HEA-340 in steel quality S355.

Self-weight of steel beams is calculated automatically from structural model, slabs are subjected to the following actions:

$g_s = 2.11 kN/m^2$ self weight of composite slab
$g_p = 2.50 kN/m^2$ permanent load
$p = 4.00 kN/m^2$ imposed load – category C2
$s = 1.21 kN/m^2$ snow load

In this example we will calculate the critical temperature and actual temperature of typical structural elements. In the case that structural element cannot provide required fire resistance, fire protection calculation will be performed.

Fire Resistance of Primary Beams

Primary beams are simply supported by columns and carry secondary beams of composite slab. They are constructed from IPE-360 profile of steel grade S275.

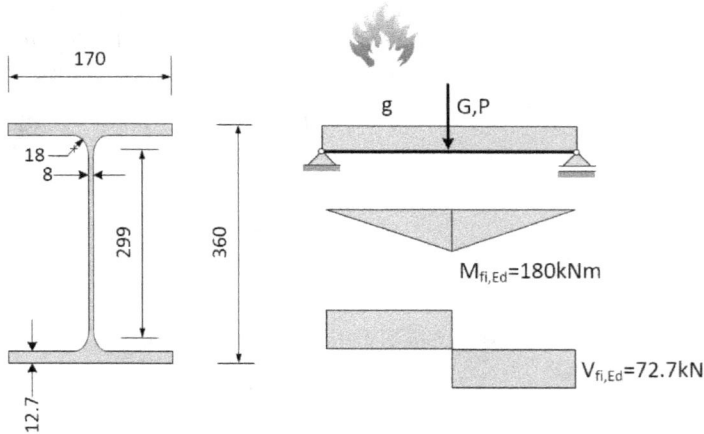

Fig. 6.2 Primary beam of cross section IPE-360 subjected to bending

According to EN 1990 fire load combination is displayed as

$$E_{fi,d,t} = \sum_{j \geq 1} G_{k,j} + \psi_{1,1} Q_P + \psi_2 Q_S = \sum_{j \geq 1} G_{k,j} + 0.7 Q_P + 0 Q_S$$

Reduced value of the parameter of section classification for steel grade S275

$$\varepsilon = 0.85 \sqrt{\frac{235}{f_y}} = 0.85 \sqrt{\frac{235}{275}} = 0.79$$

The class of the web in bending

$$\frac{c}{t_w} = \frac{29.9}{0.80} = 37.30 < 72\varepsilon = 56.88 \rightarrow Class\ 1$$

The class of the flange in compression

$$\frac{c}{t_f} = \frac{6.30}{1.27} = 4.96 < 9\varepsilon = 7.1 \rightarrow Class\ 1$$

The cross section is Class 1!

Plastic section modulus, shear area and yield strength of steel

$$W_{pl} = 1019.0cm^3 \quad A_v = 35.14cm^2 \quad f_y = 275\frac{N}{mm^2}$$

Design moment resistance of a member at room temperature

$$M_{Rd} = \frac{W_{pl}\ f_y}{\gamma_{M,0}} = \frac{1019.0 * 275}{1.0 * 1000} = 280.22kNm$$

Design shear resistance of a member at room temperature

$$V_{Rd} = \frac{A_v\ f_y}{\sqrt{3}\gamma_{M,0}} = \frac{35.14 * 275}{\sqrt{3} * 1.0 * 10} = 557.92kNm$$

Degree of utilization

$$\mu_{0,M} = \frac{M_{fi,Ed}}{M_{Rd}} = \frac{180}{280.22} = 0.642 \geq 0.013$$

$$\mu_{0,V} = \frac{V_{fi,Ed}}{V_{Rd}} = \frac{72.7}{557.92} = 0.13 \geq 0.013$$

$$\mu_0 = max(\mu_{0,M}, \mu_{0,V}) = 0.642$$

The critical temperature can be calculated by the following expression

$$\vartheta_{d,cr} = 39.19\ ln\left[\frac{1}{0.9674\mu_0^{3.833}} - 1\right] + 482$$
$$= 39.19\ ln\left[\frac{1}{0.9674 * 0.642^{3.833}} - 1\right] + 482$$
$$= 542.13°C$$

Calculation of the temperature of unprotected profile after 30min of standard fire exposure on all four sides.

At the starting moment t=0s, the steel temperature is equal to the room temperature and we have

$$\vartheta_m = 20°C, t = 0s$$

The section factor, box value of the section factor and the correction factor for the shadow effect are calculated as follows

$$\frac{A_m}{V} = 186.25 \ m^{-1} \quad \left[\frac{A_m}{V}\right]_b = \frac{106}{72.7} * 100 = 145.8 \ m^{-1}$$

$$k_{sh} = 0.9 * \frac{145.8}{186.25} = 0.70$$

First time interval t_1 can be taken as follows

$$\Delta t = 5s \longrightarrow t_1 = 0 + \Delta t = 5s$$

Gas temperature of the fire compartment is calculated using the standard temperature-time curve

$$\vartheta_g = 20 + 345 \log_{10}\left(\frac{8t_1}{60} + 1\right) = 96.54°C$$

Specific heat of steel is temperature dependent and is calculated to EN 1993-1-2(3.4.1.2) as follows

$$c_a = 425 + 0.773\vartheta_m - 0.00169\vartheta_m^2 + 0.00000\vartheta_m^3 = 439.8 \ J/kgK$$

Net heat flux can be calculated to EN 1991-1-2(3.1)

$$h_{net,d} = \alpha_c(\vartheta_g - \vartheta_m) + \Phi\varepsilon_m\varepsilon_f\sigma[(\vartheta_r + 273)^4 - (\vartheta_m + 273)^4]$$
$$= 2361.07 \ W/m^2$$

Using EN 1991-1-2(3.2.1, 3.1(6, 7)) we have the following parameters used in the formula above

$$\alpha_c = 25 \ W/m^2 \ K \quad \varepsilon_m = 0.7 \quad \varepsilon_f = 1.0 \quad \Phi = 1.0$$

$$\sigma = 5.67 * 10^{-8} \ W/m^2 \ K^4$$

For fully fire engulfed steel members we can take

$$\vartheta_{r=}\vartheta_g$$

The increase in steel temperature of the member in the time interval can be calculated as follows

$$\Delta\vartheta_{a,t} = k_{sh} \frac{\frac{A_m}{V}}{c_a \rho_a} h_{net,d} \, \Delta t = 0.70 * \frac{186.25}{439.8 * 7850} * 2361.1 * 5.0$$
$$= 0.449°C$$

Temperature in the first interval can be calculated as follows

$$\vartheta_m = \vartheta_m + \Delta\vartheta_{a,t} = 20 + 0.449 = 20.449°C$$

The next interval can be defined as

$$t_2 = t_1 + \Delta t = 5 + 5 = 10s$$

Now we can calculate the increase in temperature of the steel member in the next interval as follows

$$\vartheta_g = 146.95°C \quad c_a = 440.12 \, J/kgK \quad h_{net,d} = 4102.74W/m^2$$

$$\Delta\vartheta_{a,t} = 0.779°C \quad \vartheta_m = 20.449 + 0.779 = 21.23°C$$

The next interval can be defined as

$$t_i = t_{i-1} + \Delta t = 10 + 5 = 15s$$

The other iterations of calculation are presented in the following table:

t [min]	5	10	15	20	25	30
$\vartheta_g[°C]$	576.41	678.43	738.56	781.30	814.60	841.80
$\vartheta_a[°C]$	219.77	462.10	624.50	713.20	744.75	803.75

Temperature of steel section after 30min of fire exposure is 803.75 °C, critical temperature is reached after 12.17min – so **fire protection is needed** to achieve required fire protection **R30**.

Now we need to calculate the temperature of protected profile after 30min of standard fire exposure on all four sides.

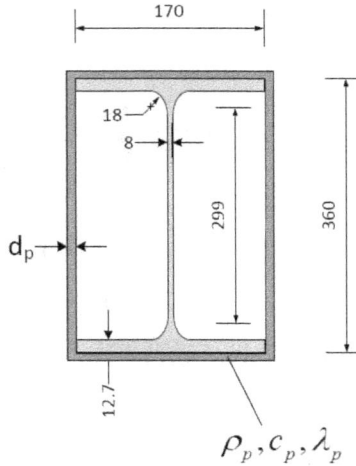

Fig. 6.3 Protected steel section of cross section IPE-360

Steel member is protected with fibre-silicate boards with the following parameters

$$\rho_p = 600\,\frac{kg}{m^3} \quad c_p = 1200\,\frac{J}{kgK} \quad \lambda_p = \frac{0.15W}{mK} \quad d_p = 0.02m$$

The section factor and the unit mass of steel are calculated as follows

$$\frac{A_p}{V} = 145.8\ m^{-1} \quad \rho_a = 7850\,\frac{kg}{m^3}$$

First time interval t_1 can be taken as follows

$$\Delta t = 30s \longrightarrow t_1 = 0 + \Delta t = 30s$$

Gas temperature of the fire compartment is calculated using the standard temperature-time curve

$$\vartheta_{g,t} = 20 + 345 \log_{10}\left(\frac{8t_1}{60} + 1\right) = 261.14°C$$

The increase in gas temperature of the fire compartment is calculated as

$$\Delta\vartheta_{g,t} = 261.14 - 20 = 241.14°C$$

Specific heat of steel is temperature dependent and is calculated to EN 1993-1-2 (3.4.1.2) as follows

$$\vartheta_{a,t} = 20°C \ \rightarrow c_a = 425 + 0.773\vartheta_{a,t} - 0.00169\vartheta_{a,t}^2 + 0.00000\vartheta_{a,t}^3$$
$$= 439.8\,J/kgK$$

The amount of heat stored in the insulation is calculated as

$$\Phi = \frac{c_p d_p \rho_p}{c_a \rho_a} \frac{A_p}{V} = \frac{1200 * 600 * 0.02}{439.8 * 7850} * 145.8 = 0.61$$

The increase in steel temperature of the member in the time interval can be calculated as follows

$$\Delta\vartheta_{a,t} = \frac{\lambda_p A_p/V \left(\vartheta_{g,t} - \vartheta_{a,t}\right)}{d_p c_a \rho_a (1 + \Phi/3)} \Delta t - \left(e^{\frac{\Phi}{10}} - 1\right)\Delta\vartheta_{g,t}$$

$$= \frac{0.15 * 145.8 * 241.14 * 30}{0.02 * 439.8 * 7850 * \left(1 + \frac{0.61}{3}\right)} - \left(e^{\frac{0.61}{10}} - 1\right)$$

$$* 241.14 = -13.2°C$$

As negative increments of steel temperature are not allowed, we have

$$\Delta\vartheta_{g,t} = 241.14°C \geq 0 \quad \rightarrow \quad \Delta\vartheta_{a,t} = 0°C$$

Second time interval t_2 can be taken as follows

$$t_2 = t_1 + \Delta t = 30 + 30 = 60s$$

Gas temperature of the fire compartment is calculated using the standard temperature-time curve

$$\vartheta_{g,t} = 20 + 345 \log_{10}\left(\frac{8t_2}{60} + 1\right) = 349.21°C$$

The increase in gas temperature of the fire compartment is calculated as

$$\Delta\vartheta_{g,t} = 349.21 - 261.14 = 88.07°C$$

Specific heat of steel and the amount of heat stored in the insulation are the same as in previous iteration

$$\vartheta_{a,t} = 20°C \quad \rightarrow \quad c_a = 425 + 0.773\vartheta_{a,t} - 0.00169\vartheta_{a,t}^2 + 0.00000\vartheta_{a,t}^3$$

$$= 439.8 \frac{J}{kgK} \quad \Phi = 0.61$$

The increase in steel temperature of the member in the time interval can be calculated as follows

$$
\Delta\vartheta_{a,t} = \frac{\lambda_p A_p/V\left(\vartheta_{g,t} - \vartheta_{a,t}\right)}{d_p c_a \rho_a (1 + \Phi/3)} \Delta t - \left(e^{\frac{\Phi}{10}} - 1\right)\Delta\vartheta_{g,t}
$$

$$
= \frac{0.15 * 145.8 * 329.21 * 30}{0.02 * 439.8 * 7850 * \left(1 + \dfrac{0.61}{3}\right)} - \left(e^{\frac{0.61}{10}} - 1\right)
$$

$$
* 88.07 = -2.92°C
$$

The other iterations of calculation are presented in the following table:

t [min]	5	10	15	20	25	30
$\vartheta_g[°C]$	576.41	678.42	738.56	781.35	814.60	841.80
$\vartheta_a[°C]$	36.33	74.64	116.40	158.33	199.30	238.85

Temperature of the steel section after 30 minutes of standard fire exposure is 238.85 degrees Celsius.

Primary beams fulfill standard fire resistance requirement R30 as it is

$$
\vartheta_a = 238.85°C < \vartheta_{d,cr} = 542.13°C
$$

Bottom Line

First critical temperature was calculated as 542.13 °C, then we calculated the development of the temperature in unprotected steel section. It was obvious that the critical temperature was reached after 12.17 min of standard fire exposure. So we realized, that steel section should be protected with insulation to fulfill the required fire protection R30. Fire protection with fibre-silicate boards was selected and after calculation of the development of the temperature in protected steel section, it was shown, that after 30min, temperature in steel section is less then critical temperature of the section, so the conclusion is that fire protected steel element fulfills the standard fire resistance requirement of R30.

Fire Resistance of Secondary Beams

Secondary beams are simply supported by main beams and support composite slab. They are constructed from IPE-330 profile of steel grade S275.

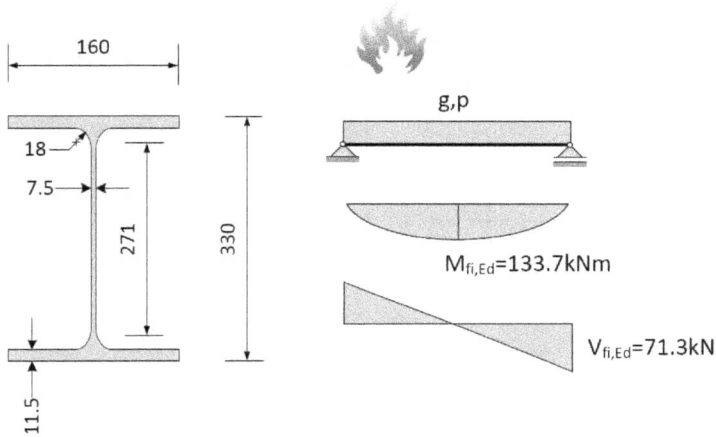

Fig. 6.4 Secondary beam of cross section IPE-330 subjected to bending

According to EN 1990 fire load combination is displayed as

$$E_{fi,d,t} = \sum_{j\geq1} G_{k,j} + \psi_{1,1} Q_P + \psi_2 Q_S = \sum_{j\geq1} G_{k,j} + 0.7 Q_P + 0 Q_S$$

Reduced value of the parameter of section classification for steel grade S275

$$\varepsilon = 0.85 \sqrt{\frac{235}{f_y}} = 0.85 \sqrt{\frac{235}{275}} = 0.79$$

The class of the web in bending

$$\frac{c}{t_w} = \frac{27.1}{0.75} = 36.10 < 72\varepsilon = 56.88 \rightarrow Class\ 1$$

The class of the flange in compression

$$\frac{c}{t_f} = \frac{5.80}{1.15} = 5.04 < 9\varepsilon = 7.1 \rightarrow Class\ 1$$

The cross section is Class 1!

Plastic section modulus, shear area and yield strength of steel

$$W_{pl} = 804.3 cm^3 \quad A_v = 30.80 cm^2 \quad f_y = 275 \frac{N}{mm^2}$$

Design moment resistance of a member at room temperature

$$M_{Rd} = \frac{W_{pl} f_y}{\gamma_{M,0}} = \frac{804.3 * 275}{1.0 * 1000} = 221.18 kNm$$

Design shear resistance of a member at room temperature

$$V_{Rd} = \frac{A_v f_y}{\sqrt{3}\gamma_{M,0}} = \frac{30.80 * 275}{\sqrt{3} * 1.0 * 10} = 489.02 kNm$$

Degree of utilization

$$\mu_{0,M} = \frac{M_{fi,Ed}}{M_{Rd}} = \frac{133.7}{221.18} = 0.60 \geq 0.013$$

$$\mu_{0,V} = \frac{V_{fi,Ed}}{V_{Rd}} = \frac{71.3}{489.02} = 0.15 \geq 0.013$$

$$\mu_0 = max(\mu_{0,M}, \mu_{0,V}) = 0.60$$

The critical temperature can be calculated by the following expression

$$\vartheta_{d,cr} = 39.19 \ln\left[\frac{1}{0.9674\mu_0^{3.833}} - 1\right] + 482$$

$$= 39.19 \ln\left[\frac{1}{0.9674 * 0.60^{3.833}} - 1\right] + 482 = 552.98°C$$

Calculation of the temperature of unprotected profile after 30min of standard fire exposure on all four sides.

At the starting moment t=0s, the steel temperature is equal to the room temperature and we have

$$\vartheta_m = 20°C, t = 0s$$

The section factor, box value of the section factor and the correction factor for the shadow effect are calculated as follows

$$\frac{A_m}{V} = 200.17 \, m^{-1} \quad \left[\frac{A_m}{V}\right]_b = \frac{98}{62.6} * 100 = 156.5 \, m^{-1}$$

$$k_{sh} = 0.9 * \frac{156.5}{200.17} = 0.70$$

First time interval t_1 can be taken as follows

$$\Delta t = 5s \longrightarrow t_1 = 0 + \Delta t = 5s$$

Gas temperature of the fire compartment is calculated using the standard temperature-time curve

$$\vartheta_g = 20 + 345 \log_{10}\left(\frac{8t_1}{60} + 1\right) = 96.54°C$$

Specific heat of steel is temperature dependent and is calculated to EN 1993-1-2(3.4.1.2) as follows

$$c_a = 425 + 0.773\vartheta_m - 0.00169\vartheta_m^2 + 0.000000\vartheta_m^3 = 439.8\,J/kgK$$

Net heat flux can be calculated to EN 1991-1-2(3.1)

$$h_{net,d} = \alpha_c(\vartheta_g - \vartheta_m) + \Phi\varepsilon_m\varepsilon_f\sigma[(\vartheta_r + 273)^4 - (\vartheta_m + 273)^4]$$
$$= 2361.07\,W/m^2$$

Using EN 1991-1-2(3.2.1, 3.1(6, 7)) we have the following parameters used in the formula above

$$\alpha_c = 25\,W/m^2\,K \quad \varepsilon_m = 0.7 \quad \varepsilon_f = 1.0 \quad \Phi = 1.0$$

$$\sigma = 5.67 * 10^{-8}\,W/m^2\,K^4$$

For fully fire engulfed steel members we can take

$$\vartheta_{r=}\vartheta_g$$

The increase in steel temperature of the member in the time interval can be calculated as follows

$$\Delta\vartheta_{a,t} = k_{sh}\frac{\frac{A_m}{V}}{c_a\rho_a}h_{net,d}\,\Delta t = 0.70 * \frac{200.17}{439.8 * 7850} * 2361.1 * 5.0$$
$$= 0.482°C$$

Temperature in the first interval can be calculated as follows

$$\vartheta_m = \vartheta_m + \Delta\vartheta_{a,t} = 20 + 0.482 = 20.482°C$$

The next interval can be defined as

$$t_2 = t_1 + \Delta t = 5 + 5 = 10s$$

Now we can calculate the increase in temperature of the steel member in the next interval as follows

$$\vartheta_g = 146.95°C \quad c_a = 440.14\,J/kgK \quad h_{net,d} = 4101.78W/m^2$$

$$\Delta\vartheta_{a,t} = 0.836°C \quad \vartheta_m = 20.482 + 0.836 = 21.32°C$$

The next interval can be defined as

$$t_i = t_{i-1} + \Delta t = 10 + 5 = 15s$$

The other iterations of calculation are presented in the following table:

t [min]	5	10	15	20	25	30
$\vartheta_g[°C]$	576.41	678.43	738.56	781.30	814.60	841.80
$\vartheta_a[°C]$	231.35	479.30	637.24	719.24	750.19	**810.48**

Temperature of steel section after 30min of fire exposure is 810.48 degrees Celsius, critical temperature is reached after 12min – **so fire protection is needed** to achieve required fire protection **R30**.

Now we need to calculate the temperature of protected profile after 30min of standard fire exposure on all four sides.

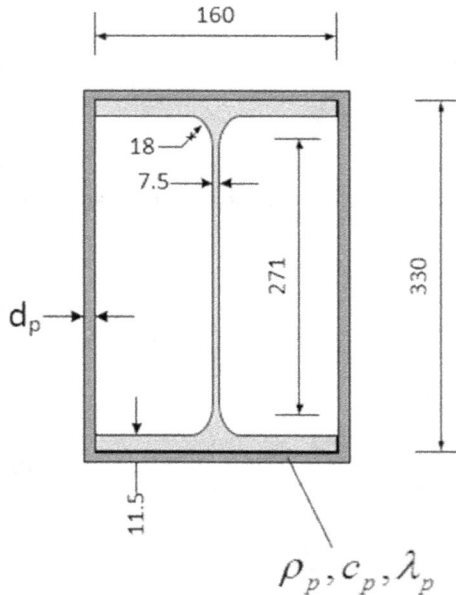

Fig. 6.5 Protected steel section of cross section IPE-330

Steel member is protected with fibre-silicate boards with the following parameters

$$\rho_p = 600 \frac{kg}{m^3} \quad c_p = 1200 \frac{J}{kgK} \quad \lambda_p = \frac{0.15W}{mK} \quad d_p = 0.02m$$

The section factor and the unit mass of steel are calculated as follows

$$\frac{A_p}{V} = 156.55 \; m^{-1} \quad \rho_a = 7850 \frac{kg}{m^3}$$

First time interval t_1 can be taken as follows

$$\Delta t = 30s \longrightarrow t_1 = 0 + \Delta t = 30s$$

Gas temperature of the fire compartment is calculated using the standard temperature-time curve

$$\vartheta_{g,t} = 20 + 345 \log_{10} \left(\frac{8t_1}{60} + 1 \right) = 261.14°C$$

The increase in gas temperature of the fire compartment is calculated as

$$\Delta\vartheta_{g,t} = 261.14 - 20 = 241.14°C$$

Specific heat of steel is temperature dependent and is calculated to EN 1993-1-2(3.4.1.2) as follows

$$\vartheta_{a,t} = 20°C \;\; \rightarrow c_a = 425 + 0.773\vartheta_{a,t} - 0.00169\vartheta_{a,t}^2 + 0.00000\vartheta_{a,t}^3$$
$$= 439.8 \, J/kgK$$

The amount of heat stored in the insulation is calculated as

$$\Phi = \frac{c_p d_p \rho_p}{c_a \rho_a} \frac{A_p}{V} = \frac{1200 * 600 * 0.02}{439.8 * 7850} * 156.55 = 0.653$$

The increase in steel temperature of the member in the time interval can be calculated as follows

$$\Delta\vartheta_{a,t} = \frac{\lambda_p A_p/V(\vartheta_{g,t} - \vartheta_{a,t})}{d_p c_a \rho_a(1 + \Phi/3)} \Delta t - \left(e^{\frac{\Phi}{10}} - 1 \right) \Delta\vartheta_{g,t}$$
$$= \frac{0.15 * 156.55 * 241.14 * 30}{0.02 * 439.8 * 7850 * \left(1 + \frac{0.653}{3}\right)} - \left(e^{\frac{0.653}{10}} - 1 \right)$$
$$* 241.14 = -14.25°C$$

As negative increments of steel temperature are not allowed, we have

$$\Delta\vartheta_{g,t} = 241.14°C \geq 0 \;\; \rightarrow \;\; \Delta\vartheta_{a,t} = 0°C$$

Second time interval t_2 can be taken as follows

$$t_2 = t_1 + \Delta t = 30 + 30 = 60s$$

Gas temperature of the fire compartment is calculated using the standard temperature-time curve

$$\vartheta_{g,t} = 20 + 345 \log_{10}\left(\frac{8t_2}{60} + 1\right) = 349.21°C$$

The increase in gas temperature of the fire compartment is calculated as

$$\Delta\vartheta_{g,t} = 349.21 - 261.14 = 88.07°C$$

Specific heat of steel and the amount of heat stored in the insulation are the same as in previous iteration

$$\vartheta_{a,t} = 20°C \quad \rightarrow \quad c_a = 425 + 0.773\vartheta_{a,t} - 0.00169\vartheta_{a,t}^2 + 0.0000\vartheta_{a,t}^3$$

$$= 439.8\frac{J}{kgK} \qquad \Phi = 0.653$$

The increase in steel temperature of the member in the time interval can be calculated as follows

$$\Delta\vartheta_{a,t} = \frac{\lambda_p A_p/V(\vartheta_{g,t} - \vartheta_{a,t})}{d_p c_a \rho_a (1 + \Phi/3)}\Delta t - \left(e^{\frac{\Phi}{10}} - 1\right)\Delta\vartheta_{g,t}$$

$$= \frac{0.15 * 156.55 * 329.21 * 30}{0.02 * 439.8 * 7850 * \left(1 + \frac{0.653}{3}\right)} - \left(e^{\frac{0.653}{10}} - 1\right)$$

$$* 88.07 = -3.18°C$$

The other iterations of calculation are presented in the following table:

t [min]	5	10	15	20	25	30
$\vartheta_g[°C]$	576.41	678.42	738.56	781.35	814.60	841.80
$\vartheta_a[°C]$	37.14	77.49	121.36	165.20	207.91	249.10

Temperature of the steel section after 30 minutes of standard fire exposure is 249.10 degrees Celsius.

Secondary beams fulfill standard fire resistance requirement R30 as it is

$$\vartheta_a = 249.10°C < \vartheta_{d,cr} = 552.98°C$$

Bottom Line

First critical temperature was calculated as 552.98 °C, then we calculated the development of the temperature in unprotected steel section. It was obvious that the critical temperature was reached after 12 min of standard fire exposure. So we realized, that steel section should be protected with insulation to fulfill the required fire protection R30. Fire protection with fibre-silicate boards was selected and after calculation of the development of the temperature in protected steel section, it was shown, that after 30min, temperature in steel section is less then critical temperature of the section, so the conclusion is that fire protected steel element fulfills the standard fire resistance requirement of R30.

Fire Resistance of Columns

Colums are part of the frame structure and are embeded into spot footings. They are constructed from HEA-340 profile of steel grade S355.

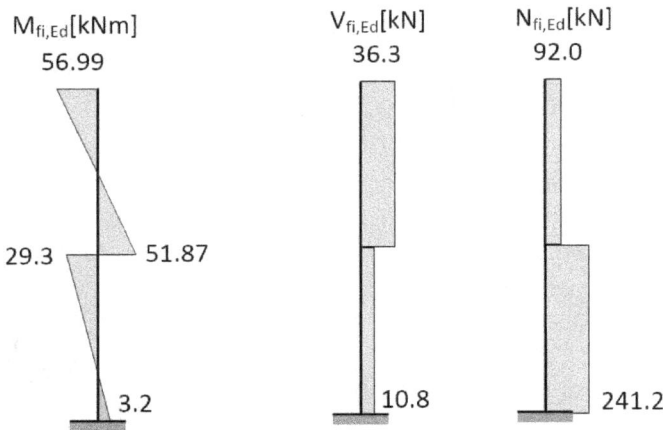

Fig. 6.6 Action effects of steel column

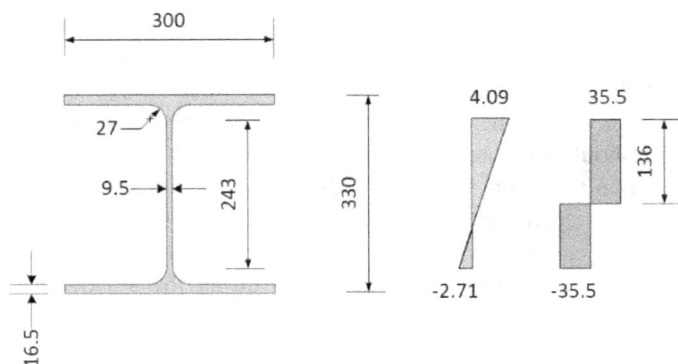

Fig. 6.7 Cross section dimensions including elastic and plastic stress diagrams for unprotected IPE-330 profile

According to EN 1990 fire load combination is displayed as

$$E_{fi,d,t} = \sum_{j \geq 1} G_{k,j} + \psi_{1,1} Q_P + \psi_2 Q_S = \sum_{j \geq 1} G_{k,j} + 0.7\, Q_P + 0Q_S$$

Stress coefficients from diagrams can be calculated as follows

$$\psi = \frac{-2.71}{4.09} = -0.66$$

$$\alpha = \frac{1}{c}\left[\frac{h}{2} + \frac{1}{2}\frac{N_{Ed}}{t_w f_y} - (t_f + r)\right]^{[11]}$$

$$= \frac{1}{24.3}\left[\frac{33}{2} + \frac{1}{2}\frac{92*10}{0.95*355} - (1.65 + 2.7)\right] = 0.56$$

Reduced value of the parameter of section classification for steel grade S355

$$\varepsilon = 0.85\sqrt{\frac{235}{f_y}} = 0.85\sqrt{\frac{235}{355}} = 0.69$$

Web in combined bending and compression

$$\frac{c}{t_w} = 25.58 < \frac{42*0.69}{0.67 - 0.33*0.66} = 64.09 \rightarrow Class\ 3\ or\ better$$

$$\frac{c}{t_w} = 25.58 < \frac{396*0.69}{13*0.56 - 1} = 43.5 \rightarrow Class\ 1$$

The class of the flange in compression

$$\frac{c}{t_f} = 7.17 < 14\varepsilon = 9.66 \rightarrow Class\ 3$$

The cross section is Class 3!

Cross section area, elastic section modulus, governing radiuses of gyration and yield strength of steel are as follows

$$A = 133.5\ cm^2 \quad W_{y,el} = 1678cm^3 \quad i_y = 14.4cm$$

$$i_z = 7.5cm \quad f_y = 355\frac{N}{mm^2}$$

Column is part of a braced structure so buckling length is taken as 0.7 of the system length in strong axis and system length in the weak axis of the column.

Non-dimensional slenderness at room temperature is calculated as follows

$$\varepsilon = \sqrt{\frac{235}{355}} = 0.81 \quad \lambda_1 = 93.9\varepsilon = 76.4$$

$$\lambda_y = \frac{l_{fi}}{i_y} = \frac{210}{14.4} = 14.58 \quad \rightarrow \bar{\lambda}_y = \frac{\lambda_y}{\lambda_1} = \frac{14.58}{76.4} = 0.19$$

$$\lambda_z = \frac{l_{fi}}{i_z} = \frac{300}{7.50} = 40.00 \quad \rightarrow \bar{\lambda}_z = \frac{\lambda_z}{\lambda_1} = \frac{40.0}{76.4} = 0.53$$

Non-dimensional slenderness at elevated temperature is calculated as follows

$$\bar{\lambda}_{\vartheta=20°C,y} = \bar{\lambda}_y = 0.19 \quad \bar{\lambda}_{\vartheta=20°C,z} = \bar{\lambda}_z = 0.53$$

Reduction factors for flexural buckling are calculated as follows

$$\alpha = 0.65\sqrt{\frac{235}{355}} = 0.53$$

$$\varphi_{\vartheta,y} = \frac{1}{2}\left(1 + \alpha\bar{\lambda}_{\vartheta,y} + \bar{\lambda}_{\vartheta,y}^2\right) = \frac{1}{2}(1 + 0.53 * 0.19 + 0.19^2) = 0.57$$

$$\varphi_{\vartheta,z} = \frac{1}{2}\left(1 + \alpha\bar{\lambda}_{\vartheta,z} + \bar{\lambda}_{\vartheta,z}^2\right) = \frac{1}{2}(1 + 0.53 * 0.53 + 0.53^2) = 0.78$$

$$\chi_{fi,y} = \frac{1}{\varphi_{\vartheta,y} + \sqrt{\varphi_{\vartheta,y}^2 - \bar{\lambda}_{\vartheta,y}^2}} = \frac{1}{0.57 + \sqrt{0.57^2 - 0.19^2}} = 0.90$$

$$\chi_{fi,z} = \cfrac{1}{\varphi_{\vartheta,z} + \sqrt{\varphi_{\vartheta,z}^2 - \bar{\lambda}_{\vartheta,z}^2}} = \cfrac{1}{0.78 + \sqrt{0.78^2 - 0.53^2}} = 0.74$$

Elastic critical moment for LTB

$$M_{cr} = 8234.16 kNm$$

Non-dimensional slenderness for LTB

$$\bar{\lambda}_{LT} = \sqrt{\frac{W_y f_y}{M_{cr}}} = \sqrt{\frac{1678 * 355}{8234.16 * 10^3}} = 0.27$$

For uniform moment factors check EN 1993-1-2 (4.2.3.5) Figure 4.2

$$\beta_{M,LT} = \beta_{M,y} = 2.44$$

Reduction factors

$$
\begin{aligned}
\mu_y &= \left(1.2\beta_{M,y} - 3\right)\bar{\lambda}_{y,\vartheta} + 0.44\beta_{M,y} - 0.29 \\
&= (1.2 * 2.44 - 3) * 0.19 + 0.44 * 2.44 - 0.29 = 0.77 \\
&\leq 0.8
\end{aligned}
$$

$$\mu_{LT} = 0.15\bar{\lambda}_{z,\vartheta}\beta_{M,LT} - 0.15 = 0.15 * 0.53 * 2.44 - 0.15 = 0.04 \leq 0.9$$

$$k_y = 1 - \cfrac{\mu_y N_{fi,Ed}}{\chi_{y,fi} A k_{y,\vartheta} \cfrac{f_y}{\gamma_{M,fi}}} = 1 - \cfrac{0.77 * 92 * 10}{0.90 * 133.5 * 1.0 \cfrac{355}{1.0}} = 0.98 \leq 3$$

$$k_{LT} = 1 - \cfrac{\mu_{LT} N_{fi,Ed}}{\chi_{z,fi} A k_{y,\vartheta} \cfrac{f_y}{\gamma_{M,fi}}} = 1 - \cfrac{0.04 * 92 * 10}{0.74 * 133.5 * 1.0 * \cfrac{355}{1.0}} = 1.00 \leq 1$$

LTB reduction factors at elevated temperature

$$
\begin{aligned}
\Phi_{LT,\vartheta,com} &= \frac{1}{2}\left(1 + \alpha\bar{\lambda}_{LT,\vartheta,com} + \bar{\lambda}_{LT,\vartheta,com}^2\right) \\
&= \frac{1}{2}(1 + 0.53 * 0.27 + 0.27^2) = 0.60
\end{aligned}
$$

$$
\begin{aligned}
\chi_{LT,fi} &= \cfrac{1}{\Phi_{LT,\vartheta,com} + \sqrt{\Phi_{LT,\vartheta,com}^2 - \bar{\lambda}_{LT,\vartheta,com}^2}} = \cfrac{1}{0.60 + \sqrt{0.60^2 - 0.27^2}} \\
&= 0.87
\end{aligned}
$$

Degree of utilization in first iteration can be calculated as

$$\mu_{0,1} = \frac{N_{fi,Ed}}{\chi_{min,fi}A\frac{f_y}{\gamma_{M,fi}}} + \frac{k_y M_{y,fi,Ed}}{W_y\frac{f_y}{\gamma_{M,fi}}}$$

$$= \frac{92 * 10}{0.74 * 133.5 * \frac{355}{1.0}} + \frac{0.98 * 56.99 * 10^3}{1678 * \frac{355}{1.0}} = 0.12$$

$$\mu_{0,1} = \frac{N_{fi,Ed}}{\chi_{z,fi}A\frac{f_y}{\gamma_{M,fi}}} + \frac{k_y M_{y,fi,Ed}}{W_y\frac{f_y}{\gamma_{M,fi}}}$$

$$= \frac{92 * 10}{0.74 * 133.5 * \frac{355}{1.0}} + \frac{1.00 * 56.99 * 10^3}{0.87 * 1678 * \frac{355}{1.0}} = 0.14$$

$$\mu_0 = max(\mu_{0,1}, \mu_{0,2}) = 0.14$$

The critical temperature in first iteration can be calculated by the following expression

$$\vartheta_{d,cr} = 39.19 \ln\left[\frac{1}{0.9674\mu_0^{3.833}} - 1\right] + 482$$

$$= 39.19 \ln\left[\frac{1}{0.9674 * 0.14^{3.833}} - 1\right] + 482 = 782.52°C$$

Because the non-dimensional slenderness is dependant on the elevated temperature, the iterative procedure is required!

We are using the value of the critical temperature 782.52 degrees Celsius from the first iteration for the calculation in the next iteration and continue with iterations until convergence is found.

The other iterations of calculation are presented in the table below.

ϑ_{cr}^i [°C]	$k_{y,\vartheta}$	$k_{E,\vartheta}$	μ_y	k_y	μ_o	ϑ_{cr}^{i+1} [°C]
782.52	0.131	0.097	0.768	0.872	0.141	777.75
777.75	0.137	0.099	0.767	0.877	0.141	777.34

The critical temperature is calculated after 2 iterations as 777.34 degrees Celsius.

Now we need to calculate the temperature of the unprotected steel profile in fire compartment after 30min of standard fire exposure on all 4 sides.

At the starting moment t=0s, the steel temperature is equal to the room temperature and we have

$$\vartheta_m = 20°C, t = 0s$$

The section factor, box value of the section factor and the correction factor for the shadow effect are calculated as follows

$$\frac{A_m}{V} = 134.41 \ m^{-1} \quad \left[\frac{A_m}{V}\right]_b = \frac{126}{133.5} * 100 = 94.38 \ m^{-1}$$

$$k_{sh} = 0.9 * \frac{134.41}{94.38} = 0.63$$

First time interval t_1 can be taken as follows

$$\Delta t = 5s \longrightarrow t_1 = 0 + \Delta t = 5s$$

Gas temperature of the fire compartment is calculated using the standard temperature-time curve

$$\vartheta_g = 20 + 345 \log_{10}\left(\frac{8t_1}{60} + 1\right) = 96.54°C$$

Specific heat of steel is temperature dependent and is calculated to EN 1993-1-2 (3.4.1.2) as follows

$$c_a = 425 + 0.773\vartheta_m - 0.00169\vartheta_m^2 + 0.00000\vartheta_m^3 = 439.8 \ J/kgK$$

Net heat flux can be calculated to EN 1991-1-2(3.1)

$$h_{net,d} = \alpha_c(\vartheta_g - \vartheta_m) + \Phi\varepsilon_m\varepsilon_f\sigma[(\vartheta_r + 273)^4 - (\vartheta_m + 273)^4]$$
$$= 2361.07 \ W/m^2$$

Using EN 1991-1-2(3.2.1, 3.1(6, 7)) we have the following parameters used in the formula above

$$\alpha_c = 25 \ W/m^2 \ K \quad \varepsilon_m = 0.7 \quad \varepsilon_f = 1.0 \quad \Phi = 1.0$$

$$\sigma = 5.67 * 10^{-8} \ W/m^2 \ K^4$$

For fully fire engulfed steel members we can take

$$\vartheta_{r=}\vartheta_g$$

The increase in steel temperature of the member in the time interval can be calculated as follows

$$\Delta\vartheta_{a,t} = k_{sh}\frac{\frac{A_m}{V}}{c_a\rho_a}h_{net,d}\,\Delta t = 0.63 * \frac{134.41}{439.8 * 7850} * 2361.1 * 5.0$$
$$= 0.29°C$$

Temperature in the first interval can be calculated as follows

$$\vartheta_m = \vartheta_m + \Delta\vartheta_{a,t} = 20 + 0.29 = 20.29°C$$

The next interval can be defined as

$$t_2 = t_1 + \Delta t = 5 + 5 = 10s$$

Now we can calculate the increase in temperature of the steel member in the next interval as follows

$$\vartheta_g = 146.95°C \quad c_a = 440.12\,J/kgK \quad h_{net,d} = 4107.33W/m^2$$

$$\Delta\vartheta_{a,t} = 0.505°C \quad \vartheta_m = 20.29 + 0.505 = 21.79°C$$

The next interval can be defined as

$$t_i = t_{i-1} + \Delta t = 10 + 5 = 15s$$

The other iterations of calculation are presented in the following table:

t [min]	5	10	15	20	25	30
ϑ_g[°C]	576.41	678.43	738.56	781.36	814.60	841.80
ϑ_a[°C]	159.25	355.32	526.50	646.79	719.95	747.34

Temperature of the steel section after 30 minutes of standard fire exposure is 747.34 degrees Celsius.

Columns fulfill standard fire resistance requirement R30 as it is

$$\vartheta_a = 747.34°C < \vartheta_{d,cr} = 777.34°C$$

Fire protection is not needed to achieve required fire resistance of **R30**.

Bottom Line

First critical temperature was calculated as 777.34°C, then we calculated the development of the temperature in unprotected steel section. It is obvious that the temperature of the steel element after 30min is less than the critical temperature of that element, so **no fire protection is needed** for achieving the required fire resistance **R30**.

Analysis of Results and Conclusions

In the table below we can see results of fire analysis for 3 typical structural elements. We can see critical temperature and the time it is reached for each of these elements.

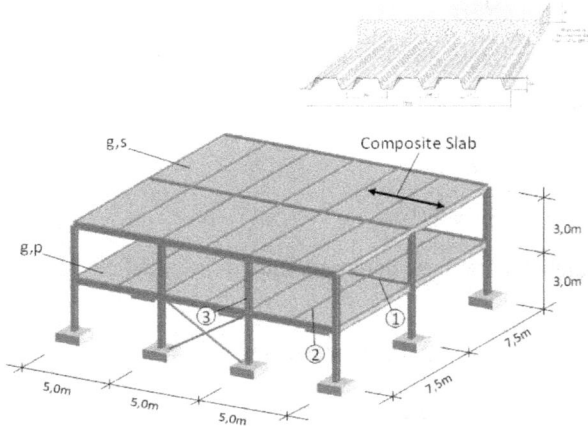

Fig. 6.8 Structure definition

Structural Element	Profile	ϑ_{cr} [°C]	R_f [min]
1- Primary beam	IPE-360	542	12.2
2- Secondary beam	IPE-330	553	12.0
3- Column	HEA-340	777	32.7

We will compare the results of fire analysis with the default value of critical temperature, which is usually used in many countries without knowing the exact value of critical temperature.

Primary and secondary beams

The calculated value of critical temperature is above default value, which means that in the case that we protect steel element with the insulation on the basis of the default values of the critical temperature, we have the insulation oversized and as such uneconomical.

Columns

The calculated value of critical temperature is far above the default value and even steel element actually does not need any insulation as its temperature in 30min is less than it's critical temperature. So if we provide insulation on the basis of the default value of the critical temperature, we have columns completely uneconomical.

Bottom Line

In order to ensure the required fire resistance requirement R30 it is necessary to provide insulation for main and secondary beams as they provide only 12 min of required fire resistance, no fire protection is required for columns.

Knowing the precise value of the critical temperature is very important because of the following:

- Savings in fire protection which can be at large structures huge
- Providing fire safety, especially for lower values of critical temperature

Knowing the exact value of Critical Temperature is cruical for fire safety optimization.

7

SOFTWARE DEVELOPMENT

I hope that you feel theoretical fundamentals presented in this handbook clear enough that you understand the theory and that worked examples served you well that you enlarged your knowledge on structural fire design.

Fire analysis presented in this handbook covers the basic principles as they are defined in Eurocodes, for the ones who want to know more there is also reference list of literature displayed at the end of this handbook.

But it is evident that using modern building design codes like Eurocodes the need to write custom software solutions is more and more real. Especially where we have complex formulas which are hardly calculated using pocket computers as it was 30 years ago. With programmable pocket computers later it was possible to write complex software, but you couldn't print the results as it is possible now. So today it is possible just by using Microsoft Excel and its programming abilities to write real software which can solve all daily engineering needs like the one we face in this handbook.

I can show you that **you do not need to be a software developer to create your own customized engineering calculations in minutes**. What is maybe the most important, you can update formulas in your calculation any time you want. This is the solution that every engineer needs, because it offers open-source solution with powerful programmable tools, but on the other side simple enough to be done instantly.

You can read other Amazon title about that
Engineering Calculations using Microsoft Excel

or you can try online course at the following address:
https://www.udemy.com/engineering-calculations-using-microsoft-excel/#/

NOTATIONS

Latin letters

A	cross section area of steel element
A_d	indirect fire actions
A_m	steel perimeter of a steel member exposed to fire
A_p	encasement perimeter of insulation exposed to fire
A_v	shear cross section of the member
A_w	cross section area of the web
A_f	cross section area of the flange
A_m/V	section factor of unprotected steel element
A_p/V	section factor of protected steel element
b	cross section width
c	straight part of web or flange of cross section
c_a	specific heat of steel
C_1, C_2	factors of the shape of the bending moment diagram
d_p	thickness of fire insulation material
E	elastic modulus of steel
$E_{fi,d}$	design action effects in fire situation
f_y	yield strength of steel at room temperature
$f_{y,\vartheta}$	yield strength of steel at elevated temperature
G	shear modulus of steel
G_k	permanent action
h	cross section height
$h_{net,d}$	design value of heat flux
I_t	torsional constant
I_y	second moment of area about strong axis
I_z	second moment of area about weak axis
I_ω	warping constant
k_1, k_2	adaptation factors for moment resistance
$k_{E,\vartheta}$	reduction factor for Young's modulus
$k_{E,\vartheta,com}$	reduction factor for Young's modulus at max. temperature
k_{sh}	correction factor for the shadow effect
$k_{y,\vartheta}$	reduction factor for yield strength
$k_{y,\vartheta,com}$	reduction factor for yield strength at max. temperature
k_z	effective length factor for lateral bending
k_ω	effective length factor for warping restraint
l_{fi}	buckling length
L	length of beam, unrestrained length of beam
$M_{b,\vartheta,Rd}$	design LTB moment resistance at elevated temperature
M_{cr}	critical moment for LTB

M_{Ed}	design bending moment in normal situation
$M_{fi,Ed}$	design bending moment in fire situation
$M_{\vartheta,Rd}$	design moment resistance at elevated temperature
$N_{b,fi,\vartheta,Rd}$	design buckling resistance at elevated temperature
N_{Ed}	design axial force in normal situation
$N_{fi,Ed}$	design axial force in fire situation
$V_{fi,Ed}$	design shear force in fire situation
$N_{fi,\vartheta,Rd}$	design axial resistance at elevated temperature
p	moisture in the insulation
P	prestressing action
$q_{fi,Ed}$	uniform load in fire situation
Q_k	variable action
r	fillet radius of cross section
$R_{fi,d,t}$	design resistance at elevated temperature
t	time
t_f	flange thickness
t_v	time delay due to the moisture in insulation
t_w	web thickness
V	cross section area of steel member
$V_{\vartheta,Rd}$	design shear resistance at elevated temperature
W_i	section modulus of the cross section
z_g	level of the application of load

Greek letters

α_c	convection heat transfer coefficient
γ_G	load factor for permanent actions in normal situation
γ_Q	load factor for variable actions at room temperature
$\gamma_{M,fi}$	material factor for fire design situation
$\gamma_{M,0}$	material factor for normal situation
ε	parameter of the section classification
ε_f	emisivity of the flame
ε_m	emisivity of the steel
η_{fi}	reduction factor for simplified rules
ϑ_d	design actual temperature of the steel element
$\vartheta_{d,cr}$	design critical temperature of the steel element
ϑ_g	gas temperature of the fire compartment
ϑ_m	temperature of the surface of the steel element
ϑ_r	radiation temperature of the fire
λ	member slenderness at room temperature
λ_p	thermal conductivity of the fire protection
$\bar{\lambda}_p$	normalized web slenderness

$\bar{\lambda}_\vartheta$	member slenderness at elevated temperature
μ_o	degree of utilization
ρ_a	unit mass of steel
ρ_p	unit mass of fire insulation
σ	Stephan Boltzman constant
χ_{fi}	reduction factor for flexural buckling
$\chi_{LT,fi}$	reduction factor for LTB
ψ_{fi}	combination factor for fire situation
Δt	time interval
$\Delta\vartheta_{a,t}$	the increase of temperature of steel element
Φ	view factor, the amount of heat stored in the fire protection

REFERENCES

1 EN 1990(2002): Eurocode – Basis of structural design, CEN(European Committee for Standardization), Brussels

2 EN 1991-1-1(2002): Eurocode 1 – Actions on structures – Part 1-1: General actions – Densities, self-weight, imposed loads for buildings, CEN (European Committee for Standardization), Brussels

3 EN 1991-1-2(2002): Eurocode 1 – Actions on structures – Part 1-2: General actions – Actions on structures exposed to fire, CEN(European Committee for Standardization), Brussels

4 EN 1991-1-3(2003): Eurocode 1 – Actions on structures – Part 1-3: General actions – Snow loads, CEN(European Committee for Standardization), Brussels

5 EN 1993-1-1(2005): Eurocode 3 – Design of steel structures – Part 1-1: General rules and rules for buildings, CEN(European Committee for Standardization), Brussels

6 EN 1993-1-2(2005): Eurocode 3 – Design of steel structures – Part 1-2: General rules – Structural fire design, CEN(European Committee for Standardization), Brussels

7 EN 1993-1-3(2006): Eurocode 3 – Design of steel structures – Part 1-3: General rules – Supplementary rules for cold-formed members and sheeting, CEN(European Committee for Standardization), Brussels

8 EN 1993-1-5(2006): Eurocode 3 – Design of steel structures – Part 1-5: General rules – Plated structural elements, CEN(European Committee for Standardization), Brussels

9 ECCS(1983): Fire Safety of Steel Structures, European Convention for Constructional Steelwork, Elsevier

10 ECCS Eurocode Design Manual – Fire Design of Steel Structures(2010), Jean-Marc Franssen, Paulo Vila Real, ECCS

11 Designers' Guide to EN 1993-1-1 Eurocode 3: Design of steel structures(2005), L.Gardner, D.A.Nethercot, Thomas Telford

www.ingramcontent.com/pod-product-compliance
Lightning Source LLC
Chambersburg PA
CBHW061610220326
41598CB00024BC/3531